NUREG/CR-5903

Validation of Smart Sensor Technologies for Instrument Calibration Reduction in Nuclear Power Plants

Manuscript Completed: December 1992
Date Published: January 1993

Prepared by
H. M. Hashemian, D. W. Mitchell, K. M. Petersen, C. S. Shell

C. E. Antonescu, NRC Project Manager

Analysis and Measurement Services Corporation
AMS 9111 Cross Park Drive, NW
Knoxville, TN 37923–4599

Prepared for
Division of Engineering
Office of Nuclear Regulatory Research
U.S. Nuclear Regulatory Commission
Washington, DC 20555
NRC FIN L2010

ISBN: 1-882148-00-2; ISBN-13: 978-1499577044; ISBN-10: 1499577044

This manuscript has been authored by a contractor of the U.S. Government under Contract No. NRC-04-91-086. Accordingly, the U.S. Government has a nonexclusive, royalty-free license to publish or reproduce the published form of this contribution, or allow others to do so, for U.S. Government purposes.

Abstract

Process instrumentation channels in nuclear power plants are calibrated at every refueling outage and surveillance tested once every month or once every quarter depending on the plant. These activities usually employ hands-on procedures which involve extensive manpower and personnel radiation exposure, and have the potential of producing maintenance-induced errors, reactor trips, and wear and tear of the plant equipment. A search of the Licensee Event Report (LER) database performed as a part of this project has shown that about 20 percent of all reactor trips and about 30 percent of all instrumentation problems in the last twelve years have been caused by hands-on maintenance and testing activities. Furthermore, a review of historical calibration and surveillance data from nuclear power plants has revealed that less than ten percent of instrument channels are normally found to drift out of tolerance over a typical fuel cycle of eighteen months. That is, more than ninety percent of the effort currently spent on surveillance testing and calibration of instrument channels may not be necessary if a means can be established to identify those instrument channels which drift or otherwise need maintenance.

Through the efforts of nuclear utilities and several research and development organizations and vendors, a number of equipment and techniques have been developed to provide early warning of incipient failures of instrument channels in nuclear power plants. These efforts have shown that an on-line monitoring system can be used to track the output of a large number of instrument channels and use routine signal processing and intercomparison techniques to identify the channels that are drifting or have bias, noise, and other problems. The on-line monitoring system can be an extension of the plant computer or a stand-alone computer. An advantage of using the plant computer is that no additional hardware such as signal isolators or analog-to-digital converters are needed, and a disadvantage is that the signals may not be available in the desired configuration and format for continuous monitoring.

An on-line monitoring system that uses the plant computer to acquire the data and a personal computer (PC) to analyze the data is currently operating at the Millstone Nuclear Power Station. A few other utilities have implemented such systems even though a formal approval has not yet been granted by the Nuclear Regulatory Commission (NRC) to any plant to switch from manual to automated calibration or surveillance testing. The NRC approval would normally require a comprehensive database of objective technical information to demonstrate the validity, accuracy and reliability of the on-line monitoring techniques and to prove that there are no common mode problems or other effects that can invalidate the tests under any circumstances. The purpose of the project reported herein is to generate such a database, and in the meantime, develop commercial equipment and techniques for the nuclear power industry. This project is performed in three phases. The Phase I effort, a nine-month feasibility study, has been completed as reported here. The Phase II effort is a comprehensive research and development project which involves theoretical work, laboratory tests, and in-plant measurements. The Phase II effort is currently underway at the Analysis and Measurement Services Corporation (AMS) and is due for completion in 1994. The Phase III project is a commercialization effort conducted concurrently with the Phase II project to provide the results of the Phase II development to interested utilities in the form of commercial test equipment, training, and engineering services.

The Phase I and Phase II projects are partially funded by the NRC under a special U.S. government program that promotes the commercialization of federally-funded research and development projects. The project is being conducted in cooperation with Duke Power Company, the host utility for the in-plant validation of the technology. The test site is the McGuire Nuclear Power Station Unit 2, a pressurized water reactor (PWR).

During Phase I, an on-line monitoring system consisting of a data acquisition cabinet and a computer was installed at the McGuire Station and connected to 170 instrument channels in the primary and secondary systems of the Unit 2 plant. The system has been monitoring the output of these instrument channels since March 1992 when the plant started its eighth fuel cycle. The in-plant work will continue until the end of this fuel cycle after which a listing of the instrument channels which have had a problem during the fuel cycle will be prepared using the on-line monitoring data. This list will be compared with the plant's surveillance and maintenance logs and the results of the hands-on calibrations to be performed during the next refueling outage. It is expected that this comparison will show, with probably some exceptions, that the on-line monitoring system can successfully identify most of the faulty instrument channels. In this case, a topical report will be prepared by AMS to be submitted to the NRC under the Duke Power Company's docket. The purpose of the topical report will be to obtain formal approval from the NRC to use the on-line monitoring system as a substitute for much of the surveillance tests and the conventional calibrations.

In addition to the in-plant tests, the Phase I project involved numerous laboratory experiments with nuclear grade sensors and instrumentation systems installed in a test loop that was designed and constructed for this project. The laboratory data were essential in Quality Assurance testing and validation of the data acquisition and data analysis algorithms and the software packages for the on-line monitoring system. In addition, the laboratory experiments provided a means to gain experience with the behavior of various instruments under various test conditions. The laboratory tests will be continued in Phase II to provide a database of information to serve as a learning tool in developing the experience that is needed to interpret the behavior of in-plant signals.

Table of Contents

List of Figures

List of Tables

1. Introduction

The nuclear power industry has been eager to implement smart sensor technologies and digital instrumentation concepts to reduce the manpower and effort currently spent on testing and calibration of process instrumentation channels in the safety and non-safety systems of pressurized and boiling water reactors. The NRC is faced with requests from nuclear utilities for approval to implement on-line monitoring techniques as a part of the justification to extend their calibration intervals from 18 months to 24 months to coincide with extended fuel cycles. Furthermore, advanced reactors currently under design and development will largely use digital instrumentation and smart sensors; the adequacy of which must be evaluated and approved by the NRC for the licensing of these reactors. In addition, plant aging concerns and life extension issues have led to extensive research in the last few years which has demonstrated that the performance of instrument channels in nuclear power plants is susceptible to aging degradation under normal use, and must therefore be tested frequently as the plant ages. These issues have stimulated the NRC to initiate the research and development project documented in this report. The project is intended to produce substantial technical data for the NRC to help in formulating the requirements that must be satisfied and the procedures that must be followed to convert from hands-on testing to automated testing if it is proven by the results of this and other similar projects that automated testing techniques are as effective as conventional techniques.

The project is divided into three phases starting with a preliminary study in Phase I to determine if a long-term study is warranted. The Phase I project has been successfully completed as reported herein and the Phase II project has been awarded. The Phase III project is a commercialization effort conducted independently by AMS. The purpose of the Phase III project is to bring the results of the Phase I and Phase II developments to the marketplace in the form of equipment, training, and services for on-line testing of calibration and response time of process instrumentation channels in nuclear power plants and other processes.

The purpose of the Phase II project is to perform a comprehensive set of experimental investigations as well as equipment and software development, software validation, verification, and Quality Assurance (QA) testing in the area of on-line verification of calibration of process instrumentation channels. These activities will be conducted using actual nuclear power plant data

as well as synthetic data generated on computers, and laboratory data generated through experiments with representative nuclear-grade sensors and signal conditioning equipment installed in a laboratory test loop. The laboratory experiments will involve inducing artificial degradation in the instruments to show that the algorithms developed in the project can successfully identify the degradation. In addition, a database of cause and effect relationships for sensor degradation tracking will be developed based on theoretical analysis and laboratory tests. The database will be used in an expert system to be developed in Phase III. The expert system, which is envisioned to be the final product of this three-phase project, will be designed to continuously monitor the output of a large number of instrument channels and automatically flag the anomalous channels and determine the cause of the anomaly. This will help determine whether the performance of the instrument channel can be restored by a new calibration, replacement of a component, or some other action is necessary. The words anomaly, degradation, and fault are used in this report synonymously to refer to any undesirable behavior in an instrument channel that is significant enough to require a maintenance action.

The work completed to date began by building a data acquisition system in Phase I using off-the-shelf hardware. The system was built and programmed to sample steady-state (DC) and noise (AC) data from any number of instrument channels at any frequency desired by the user. The DC data is used for drift monitoring and the noise data is used for determining changes in response time.

Upon construction, software validation, and QA testing, the data acquisition system developed in Phase I was installed at the McGuire Nuclear Power Station where it is continuously sampling data from 170 instrument channels in the plant's primary and secondary systems. These channels provide temperature, pressure, level, flow, and neutron flux signals to the safety and non-safety systems of the plant. The signals were hard-wired to the data acquisition unit through signal isolators. The in-plant data are stored on computer disks and subsequently analyzed off-site at AMS. The analysis is performed in two steps as follows:

1. *Data Qualification.* Each data record is screened for electrical noise, spikes, sudden shifts, and other extraneous effects. These effects are then analyzed to determine if they are due to problems within the

instrument channels. If so, they are used as diagnostic tools and their patterns and frequencies are monitored and stored. Otherwise, the affected data block(s) is discarded and the data qualification process is continued by calculating the mean, variance, skewness, and higher moments of the data. These moments are then checked to ensure that the data is not one-sided, clipped, or abnormal.

2. *Data Analysis*. The DC data are analyzed by straight and weighted averaging techniques for the redundant channels, and analytical redundancy techniques for the non-redundant channels. The analytical redundancy techniques involve empirical and physical modeling to estimate the value of a process parameter as a function of a number of other parameters. The noise data are analyzed by zero crossing, spectral pattern recognition, and autoregressive (AR) modeling techniques.

The on-line monitoring system has been operating at the McGuire Nuclear Station for nine months at this writing. Although a number of interesting phenomena has been observed, no major drift or response time degradation has been detected in any of the McGuire channels. However, a comparison of on-line monitoring results and the plant surveillance logs have shown good agreement for the few instruments that had experienced drift and were therefore adjusted by the plant's technicians to null the drift.

The McGuire plant has tripped a couple of times since the in-plant data acquisition began, providing useful heat up and cool down data to track the calibration of the instrument channels over a wide operating range.

The work at McGuire will be continued for the rest of the current fuel cycle and probably into the next fuel cycle. The current fuel cycle is due to end in June 1993. At the end of each fuel cycle, comparisons will be made between the results of the on-line monitoring tests and the hands-on calibrations. These comparisons will help determine to what extent the on-line monitoring system is successful in finding the instrument channels that have drifted out of tolerance. The level of success of the on-line monitoring system will determine the extent of relief to be expected from the NRC. It is anticipated that if the on-line monitoring results are successful, the NRC may grant an approval to the utilities who have a validated on-line monitoring system to extend their calibration intervals from 18 months to 24 months to coincide with their extended fuel cycles. In the opinion of the authors, regardless of the success of the on-line monitoring techniques, it is unlikely that the NRC would grant a permanent approval in the near future to any utility to depend solely on an on-line monitoring system to determine whether or not to calibrate an instrument channel. It is possible, however, that the use of on-line monitoring techniques combined with fault tolerant systems and new equipment with longer mean time between failures, modern predictive maintenance practices, smart sensors, and digital instrumentation and other developments may eventually lead to an NRC approval to delete much of the manual calibrations currently performed. Of course, some limit would probably be placed on this practice to ensure that every instrument is calibrated every so often (e.g., every 10 years) to reduce the risk of total dependence on new technologies until it is proven by years of experience that a total dependency on automated testing is free of any risk.

In addition to the in-plant validation tests at McGuire, the Phase I project involved laboratory work that was initiated to: 1) evaluate the methods that are available for determining the drift of instrument channels, and 2) establish the accuracy of the on-line monitoring techniques as a function of plant type, sensor type, plant conditions, redundancy, and other factors. Furthermore, as a part of the laboratory work performed in Phase I, a number of smart temperature and pressure sensors were evaluated by laboratory testing. The details of this work are presented in the body of the report. Also presented in the body of the report are the results of a search of the LER database on failures and faults of process instrumentation channels in nuclear power plants, the principles of on-line monitoring techniques, and extensive laboratory and in-plant data that were generated in the course of the Phase I project.

The Phase I project has concluded that it is simple to develop and implement an on-line monitoring system in a nuclear power plant to check for drift and response time degradation. What remains to be done, however, is to show the sensitivity, repeatability, and accuracy of such systems under a variety of plant operating conditions for a variety of sensors and instrument channels. An interesting conclusion of Phase I was that simple averaging methods and common sense analysis techniques are as effective as some of the more complicated techniques described in the literature. For response time degradation monitoring, however, simple techniques such as zero crossing were found to be only marginally useful. For this application, more sophisticated methods such as AR modeling and spectral pattern recognition must be used to provide reliable results.

It should be pointed out that several projects similar to the one described in this report have been carried out by the Electric Power Research Institute (EPRI) and other organizations in the U.S. and abroad including a few utilities.[1,2] In fact, an in-plant validation effort similar to our work at McGuire is presently underway at the San Onofre Nuclear Power Station. The Southern California Edison Company, who operates San Onofre, has made presentations to the NRC and is seeking approval from the NRC to use an on-line monitoring system as a part of the effort to be spent at San Onofre to justify extended instrument calibration intervals.[3]

It is important to point out that utilities are mainly interested in avoiding unnecessary calibrations of the sensors that are located in the containment and other hazardous areas of the plant as opposed to the rest of the instrument channels that can be conveniently calibrated while the plant is on-line. More specifically, although the calibration reduction methods described here are useful for testing a complete instrument channel, the major benefit to utilities is in avoiding the calibration of the sensors (especially pressure and differential pressure transmitters) in the field. The balance of an instrument channel is much easier to calibrate and sensors such as Resistance Temperature Detectors (RTDs) can be cross calibrated on-line during plant operation. A detailed description of the cross calibration technique is given in this report.

The information developed in the Phase I project has been used by AMS in writing a draft of the ISA Standard S67.06. This standard was issued by the Instrument Society of America (ISA) in 1984 for response time testing of instrument channels in nuclear power plants[4]. It is under revision now to include the on-line methods that have been developed since 1984 for both response time testing and calibration.

2. Background and Terminology

Following the 1979 accident at the Three Mile Island Nuclear Power Station Unit 2, the NRC implemented a number of new requirements to ensure that the reactor operators are provided with accurate, timely, and reliable information about the status of the plant under normal and accident conditions. In response, the nuclear industry began upgrading the control rooms of the plants using state-of-the-art computer technology, color monitors, and digital and analog display equipment to provide the operators with a great deal of qualitative and quantitative information at the touch of a few buttons. The displays were designed and located in the control room according to human factor principles to make it easy for the operators to determine the status of the plant at a glance. An example of an important operator aid that incorporates these new developments is the Safety Parameter Display System (SPDS) which is used to assess the safety status of the reactor and critical components of the plant. This system uses the existing signals from some of the process instrumentation channels to display the present and past status of the plant in terms of color graphs and simple charts.

To ensure that reliable signals are used in operator aids, EPRI initiated research and development activities in the early 1980s in an area that is now known as "signal validation."[5] Signal validation techniques have been used previously in the aerospace and aviation industries for flight control and space vehicle applications.

Signal validation consists of a variety of signal processing techniques implemented in nuclear power plants to ensure that sensor drift, response time degradation, bias, noise, and other sensor or system anomalies do not mislead the reactor operators. Signal validation depends on the redundancy of sensors in nuclear power plants and the physical relationships between the process parameters to check the consistency of the measurements, predict the expected values of process variables, and detect, isolate, and characterize any significant anomaly in the instrument channel.

EPRI's efforts in the signal validation area have not only produced improvements in operator aids, but also have laid the foundation for the development of on-line methods for testing the calibration of instrument channels. In fact, the outgrowth of signal validation techniques for instrument calibration testing has overshadowed its application to SPDS and other operator aids. In addition to EPRI, a number of national and international research and development organizations, universities, national laboratories, and utilities have worked in the signal validation area. As a result, numerous techniques have been researched or developed and documented under a variety of names. Some of the more common of these techniques are defined below and a detailed description is provided in Chapter 5 under the title "Smart Sensor Technologies."

Like Signal Comparison

This method is also referred to as cross calibration or DC signal comparison. It involves scanning the output of a number of instrument channels that are measuring the same process parameter and determining the deviation of each channel from the average of all channels (excluding the outliers). This method is popular in PWRs for on-line testing of calibration of temperature sensors at isothermal conditions. The principle of this method is illustrated in Figure 2.1. A problem with the like signal comparison technique is that it may not account for common mode drift unless a newly calibrated sensor is included in the comparison and/or a method such as analytical redundancy, as defined below, is used to provide an independent estimate of the process parameter.

Analytical Redundancy

This method is used when an adequate number of physically redundant channels is not available for intercomparison and when an independent measure of a process parameter is needed to account for common mode drift. As its name implies, analytical redundancy depends on theory to produce fictitious sensors or instrument channels to increase the redundancy. More specifically, it uses a group of diverse signals as the input to a physical or empirical model to produce a new signal that has a relationship with the group. Analytical redundancy is also referred to as diverse signal comparison. The principle of the method is illustrated in Figure 2.2.

Parity Space

This method is used to determine the consistency between a group of redundant signals. The common components of the signals are subtracted out and the remaining components are compared, two at a time. Based on the differences between the residual components in each pair, an inconsistency index is generated and used for diagnostics. The consistency

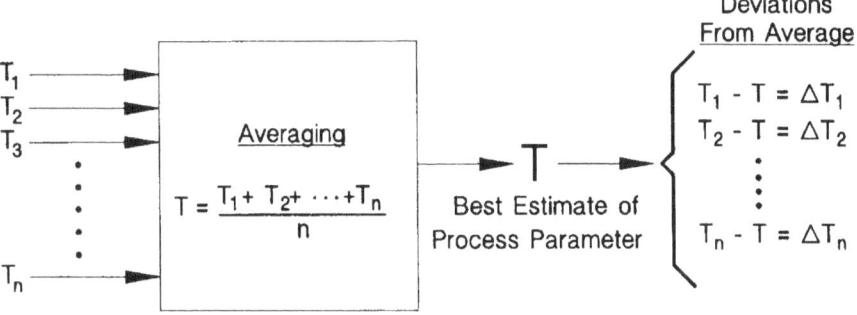

Figure 2.1 Illustration of principle of like signal comparison technique

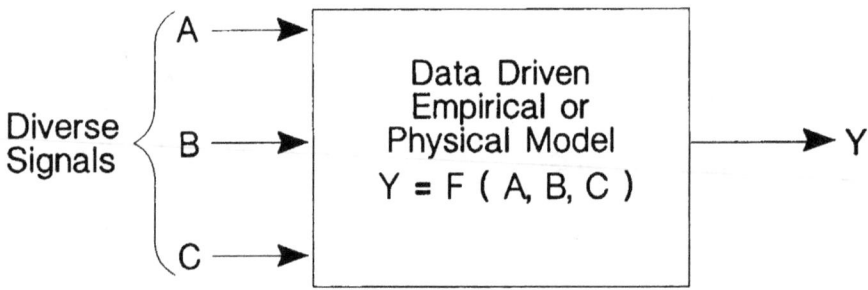

Figure 2.2   Illustration of principle of analytical redundancy

index is also used as a weighting factor in weighted averaging of redundant signals. Signals with low inconsistency indices are weighted more than signals with high inconsistency indices.

Response Time Degradation Monitoring

Response Time Degradation Monitoring uses the natural fluctuations (noise) that usually exist at the output of most instrument channels while the process is operating. The DC component of the channel output signal is first removed and the remaining signal is amplified. Any high frequency electrical noise on the signal is then filtered out and the power spectral density (PSD) of the signal is generated by a Fast Fourier Transform (FFT) or autoregressive (AR) modeling. The PSD is then presented in terms of a plot of amplitude squared versus frequency. A change in certain characteristics of a PSD would be an indication of a change in the response characteristics of the channel. Pattern recognition algorithms as well as neural networks and expert system concepts may be used for computer monitoring of changes in the shape of PSDs.

In order to distinguish sensor effects from process effects, the multivariate autoregressive (MAR) modeling approach may be used. The AR and MAR techniques are described later in this report.

Rudimentary Techniques

A number of simple operations are described in the literature on calibration reduction. These are grouped here as rudimentary techniques due to their elementary and common sense nature:

- *Limit checking*. This involves monitoring the amplitude of a signal to ensure that it remains within a predetermined band. The technique is useful for testing the plant signals for saturation and sudden changes in amplitude due to plant trips, sensor failures, etc. The data qualification techniques, which were mentioned in Chapter 1, can be used for limit checking through a calculation of the mean, variance, skewness, and higher moments of the signals. However, the data qualification techniques are more suitable for testing of AC signals and can be used for qualifying DC signals only if a long data record of DC data is available.

- *Auctioneering*. This involves rejecting the highest and lowest signal levels among three or more equivalent signals. This is similar to the approach used in the cross calibration technique where signals with certain deviations are referred to as

"outliers" and rejected from averaging when a group of signals are intercompared.

- *Instrument Loop Integrity Checking*. This is a test of an entire instrument loop from the sensor to the indicator, and it applies mainly to cables and connectors. The test generally involves measuring the electrical resistance of the whole loop in one test or testing a portion of the loop at a time. Capacitance-to-ground is another parameter that can be measured in addition to the electrical resistance. Recently, the use of time domain reflectometry (TDR) and related measurements have been growing in popularity in nuclear power plants for testing of cables and connectors in an instrument channel.[6]

- *Insulation Resistance Measurements*. Measurement of resistance to ground (Insulation Resistance) is a well-known method for monitoring the degradation of insulation material and detection of moisture in cables and instruments.

On-Line Calibration

There is currently no way to actually perform an on-line calibration on an instrument channel and obtain the same results as in a conventional hands-on calibration. Nevertheless, the term on-line calibration is sometimes used in the literature. This term refers to on-line testing of calibration to determine if the instrument's calibration is still valid and whether or not an instrument is in need of a new calibration. An alternative term is on-line calibration verification.

On-line calibration is based on the premise that detecting a change in calibration is a simpler task than an absolute calibration and a task that can be automated.

Trend Analysis

Trend analysis is often mentioned in the literature as a means of determining when the performance of an instrument or a group of instruments may reach an unacceptable level. Trend analysis may be done manually or by a computer using the instrument calibration and surveillance records. Trend analysis is highly regarded as an effective, simple, and practical means of characterizing instrument problems and defining optimum calibration and test intervals. Trend analysis is useful as a predictive maintenance tool for scheduling of outage activities ahead of time.

3. Current Nuclear Industry Practice

The nuclear power industry currently practices a very conservative approach with respect to performance testing of safety-related process instrumentation channels. In most plants, these channels are qualitatively checked three times a day, surveillance tested every month, and fully calibrated at every refueling outage and whenever a component is replaced. In addition, response time measurements are made on certain components of the instrument channels such as the sensors, filter modules, and other components that may be susceptible to response time degradation.

There are some variations in testing practices throughout the nuclear power industry and some differences in the terminologies used for the tests. For example, the monthly or quarterly surveillance tests are referred to as functional tests in some plants and are performed according to a different set of procedures and acceptance criteria than the surveillance tests. These variations make it difficult to provide a general picture of the nuclear industry's practices. Nevertheless, we have attempted to present in the following sections a representative overview of the bulk of current practices including the test frequencies and the general test procedures for the calibration and surveillance tests. The overview is preceded by a number of figures to show the typical components of an instrument channel.

Figure 3.1 is a block diagram of possible components of a typical instrument channel. This includes the sensor, signal transmission lines, signal converter and signal conditioning equipment, trip logic, and actuation modules. Figure 3.2 shows some of the specific components of an analog instrumentation channel. Recently, digital instrumentation channels have been introduced to the nuclear power industry, and a few plants have already installed and are successfully using such instruments in both the safety and non-safety systems of their plants. An example is the Westinghouse Eagle 21 system which replaces the conventional analog instrumentation systems.

3.1 Daily Channel Checks

The safety-related instrument channels in most plants are qualitatively checked by the plant operators once every shift. The operators look at the indicators in the control room to ensure that the redundant channels agree with one another within a certain tolerance. The resulting information is recorded in the plant's daily logs

and any problem to be corrected is reported to the maintenance staff.

3.2 Surveillance Tests

Surveillance tests are usually performed on all safety-related instrument channels once every month while the plant is operating. The purpose of the surveillance tests is to either verify the trip setpoints or test the functionality of the instrument channels.

The surveillance tests are performed at the instrument racks, and include all the components of the instrument channel except for the sensor. The sensor is located in the field and is not usually tested during plant operation except for in-situ response time testing described later. There is some concern as to whether or not it makes sense to test an instrument channel without the sensor. The sensor is the component of the channel that is most susceptible to performance problems because it is located in the harsh environments of the plant as opposed to the rest of the channel which is located in a mild environment. An on-line monitoring system as a substitute for the surveillance tests, as contemplated by the nuclear industry, has the advantage of testing the whole channel including the sensor. In addition, on-line monitoring is a completely passive approach in contrast with the surveillance tests which require physical interactions with the plant equipment.

Surveillance Test Procedure

The following procedure is typical for performing a surveillance test:

1. Verify that the plant conditions are suitable for the tests and no other redundant channels are tripped.

2. Place the instrument channel in "Test."

3. Disconnect the sensor from the rest of the channel. In most plants this step is performed using switches or relays on test cards provided for this purpose. There is often no need to lift leads to disconnect the sensor. Figure 3.3 shows the schematic of a typical test card and the relays that are provided to automatically disconnect the sensor from the rest of the channel.

4. Input a test signal (voltage, current, or resistance) to the instrument channel and verify that the channel

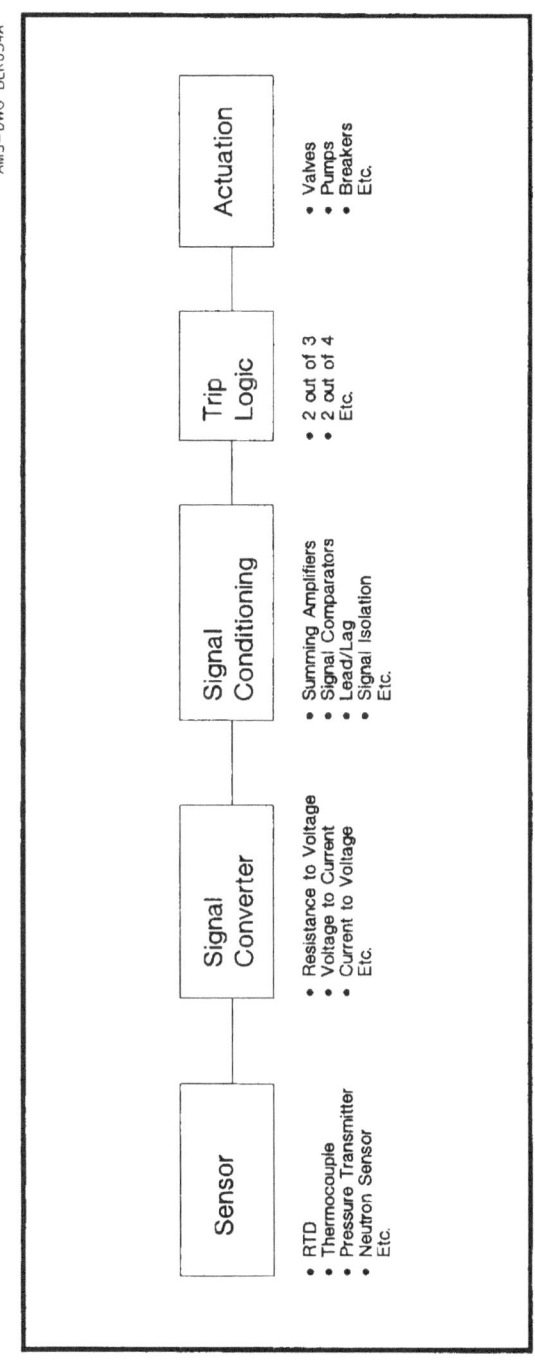

Figure 3.1 Typical components of an instrument channel in a nuclear power plant

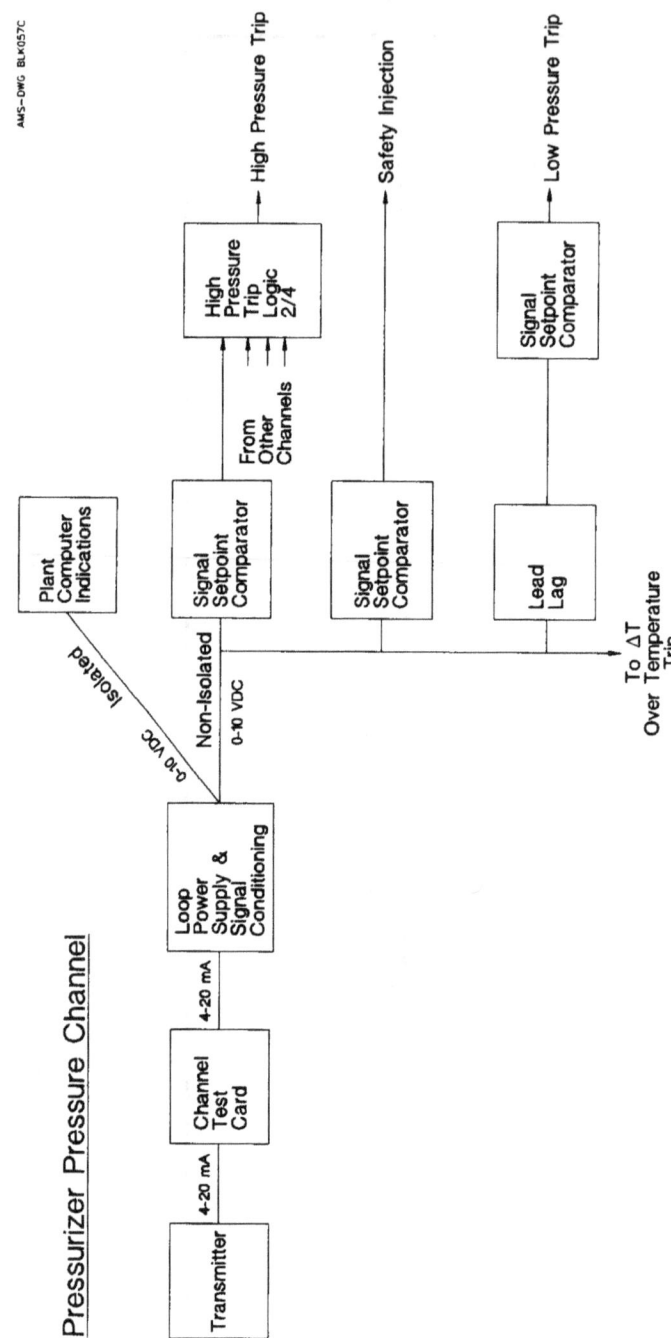

Figure 3.2 Typical components of an analog instrumentation channel

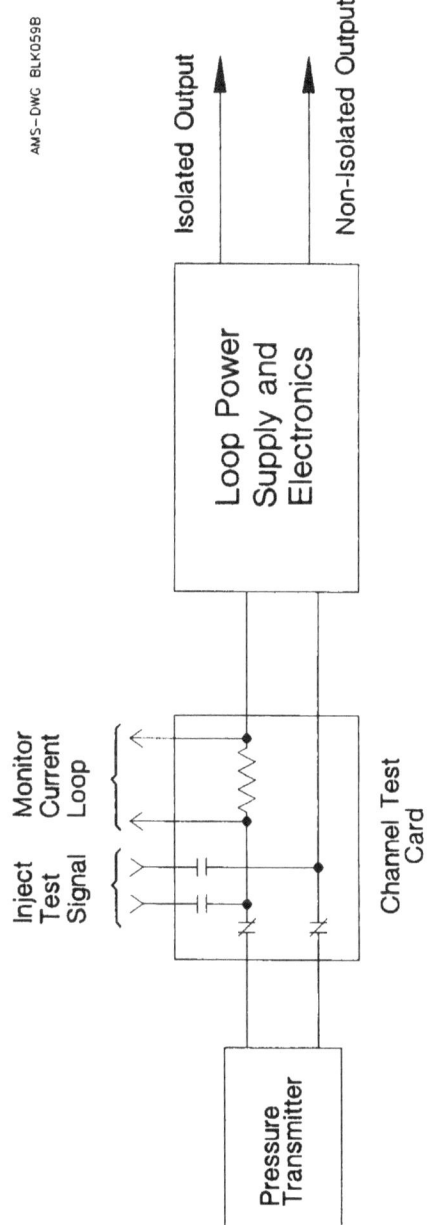

Figure 3.3   A typical pressure transmitter channel showing the channel test card

will produce a trip signal at the required setpoint. If the trip does not occur as expected, a full calibration or a bi-stable adjustment is usually performed on the channel. A typical schematic of a redundant instrument set is shown in Figure 3.4 including most of the typical components and signal paths which lead to a trip in case of an undesirable transient in the plant.

5. Return the channel to service.

The surveillance test of an instrument channel requires anywhere between a few minutes to several hours depending on the plant, the channel being tested, and the test equipment used.

Although the surveillance tests are performed monthly in a majority of the U.S. plants, there are a growing number of plants which have successfully extended their surveillance intervals to once a quarter through a combination of efforts including a comprehensive trend analysis using archived data, on-line monitoring, and a documented history of stable performance.

Surveillance Test Equipment

Surveillance test equipment generally consists of a signal generator or a power supply, a variable resistor such as a decade box, and a strip chart recorder or equivalent. To facilitate the tests, a number of modern signal generators and microprocessor-based test equipment have recently been introduced to the market or developed by utilities. As a part of the Phase I project, we conducted a number of interviews with utility personnel who had experience with such test equipment. The results were mixed and mostly negative except for one utility in the survey that has developed its own automated test unit and is using it with great success.[7]

Based on our interviews, we have concluded that while some of the automated equipment are generally helpful, they do not significantly reduce the total time spent in performing the surveillance tests. This is because a majority of time spent in a surveillance test is consumed in placing the channels in and out of test, an activity that cannot be readily automated. There has been some work spent on development of equipment that can be used for surveillance tests without having to trip the channel or disconnect the sensors. However, such systems are not currently used in any U.S. plants.

Shortcomings of Surveillance Tests

A major concern of utility technicians about surveillance tests is the increased potential for a reactor trip during the tests. When the instrument channel is placed in "Test" for surveillance, the channel is in a "tripped" state which places the reactor at a higher risk for automatic shutdown (scram). Most plants normally scram on 2 out of 4 logic meaning that they will scram when any two out of four safety-related channels exceed a setpoint. When one of the four channels is in Test, the plant will scram if only one of the remaining three channels exceeds a setpoint. Most often, a scram that occurs during the surveillance tests is caused by different test personnel working on different protection sets. The scram occurs when one channel is tripped for the surveillance tests and another channel is inadvertently tripped for the other tests. A potential remedy, although non-conservative, is to by-pass the channel instead of placing it in Test. A channel is by-passed using a constant test signal to emulate the sensor and keep the channel in its normal operating state. This approach corresponds to a 2 out of 3 logic which is less conservative than 1 out of 3 logic. Therefore, the plant technical specifications do not normally allow the test personnel to put a safety channel in by-pass.

3.3 Full-Channel Calibration

All safety-related instrument channels are fully calibrated during refueling outages. The calibration procedures are almost identical to the surveillance procedures outlined above except that they include the sensor. Furthermore, in executing calibration procedures, all instrument deviations are usually zeroed, if possible, whether or not a channel meets its acceptance criteria.

The full-channel calibration practice seems to be uniform throughout the nuclear industry except for what is done with the sensors. More specifically, the channels (excluding the sensors) are fully calibrated in all plants and all problems are resolved at every refueling outage. In addition, all safety-related pressure and differential pressure transmitters (including level and flow transmitters) are calibrated in all plants and all problems are resolved at every refueling outage. Thermocouples and neutron detectors are rarely calibrated, except for comparing neutron channel outputs to heat balance data, and the practice is sporadic with respect to RTDs. A few plants periodically remove and recalibrate their RTDs, some plants periodically install new RTDs with fresh calibrations, a majority of plants perform on-line cross calibration, and other plants do not calibrate their RTDs at all. The number of RTDs that are calibrated and the frequency of the calibration are also sporadic across the nuclear power industry.

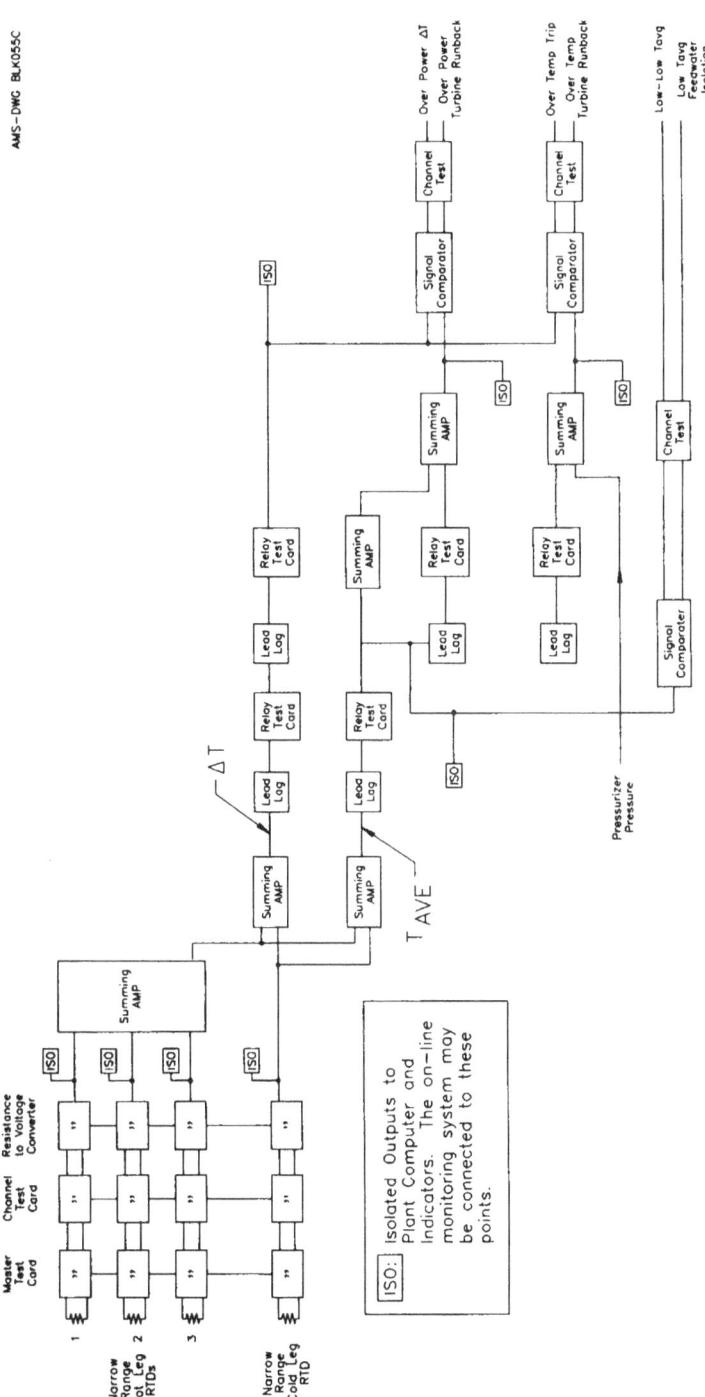

Figure 3.4 Schematic of a typical redundant instrument set in a nuclear power plant

AMS—DWG BLK055C

- 13 -

The procedures for calibration of pressure transmitters and RTDs are summarized below. The word pressure transmitter is used in the remainder of this chapter to refer to pressure, level, flow, or differential pressure transmitters.

Pressure Transmitter Calibration

A typical procedure for calibration of a pressure transmitter is as follows:

1. Isolate the transmitter from the process. This is accomplished with isolation valves already installed in the system.

2. Inject a pressure test signal to the sensor and measure its output. This is usually repeated for 0, 25, 50, 75, and 100 percent of the transmitter span. A typical calibration data sheet from a nuclear power plant procedure is shown in Table 3.1.

3. Adjust the zero and span of the sensor as necessary to bring its output within tolerance or null its deviation. In addition to zero and span, in some transmitters, a linearity adjustment is also provided, but it is rarely used in the field calibrations. The linearity adjustment is mostly used at the factory during the initial calibration of the transmitter.

4. Return the transmitter to service. In this step, isolation valves are slowly opened to avoid pressure surges when returning the transmitter to service and equalizing valves are used to equalize pressure across differential pressure transmitters to prevent overranging during isolation and return of the transmitter to service.

RTD Calibration

Several methods are available for calibration of RTDs. These methods include removal and calibration, cross calibration, and the Johnson Noise technique. Removal and calibration is the most exhaustive method and is used in only a few plants. In these plants, the RTDs are removed during a refueling outage and taken to a laboratory where they are calibrated in a constant temperature medium. This method is the most accurate and reliable way to calibrate an RTD, but it has several disadvantages. The method is time consuming and tedious, it can cause damage to the RTD, and in the case of thermowell-mounted RTDs, it can result in response time problems if the RTD is not properly seated in its thermowell after calibration. More importantly, the method involves radiation exposure to the test personnel. A better method, although not as accurate, is the cross calibration test which is used in a majority of plants performing RTD calibration.

Cross calibration tests are performed remotely from the control room area when the plant is at isothermal conditions during a scheduled shutdown or startup. The cross calibration test is based on determining the deviation of each RTD from the average of all RTDs (excluding any outliers). The advantage of the test is that it is simple and practical, and it accounts for all process conditions and installation effects on calibration. More importantly, it can be performed remotely from outside the containment. The disadvantage of the method is that it may not account for common mode calibration problems such as unidirectional drift unless one or more freshly calibrated RTDs are included every time the tests are performed. Recently, it has been shown that the drift of a large group of RTDs should be random.[8] In this case, the validity of the cross calibration method should approach that of a laboratory calibration especially if the cross calibration data is corrected for plant temperature stability and uniformity problems during the cross calibration tests.

The Johnson Noise technique has been developed for in-situ calibration of RTDs under a research and development project at the Oak Ridge National Laboratory sponsored by EPRI.[9] The method is based on measuring a voltage in the nanovolt range that corresponds to the temperature to which the RTD is exposed. Because of difficulties in measuring such a small voltage over long lengths of cables, this method in its current state of development is not accurate enough for remote calibration of RTDs in nuclear power plants. It may, however, be accurate enough if the calibrations are performed during refueling outages near the RTD in the plant, but this approach has the disadvantage of personnel radiation exposure.

3.4 Response Time Testing

Response time measurements are performed on safety-related instrument channels to ensure that the plant is shut down in a timely manner in case of an undesirable transient. The sensors are usually tested separately and their response times are added to the response time of the rest of the channel to provide the overall response time of the channel. In a few plants, however, the response time tests are performed on the whole channel including the sensor by providing a test signal such as a step change to the sensor in the field and measuring its time delay at the channel output.

Table 3.1

TRANSMITTER CALIBRATION
DATA SHEET

Duke ID	2NCFT5000/2FT-414 OUTPUT
Manuf.	BARTON
Model No.	764
Location	C1-747 TEST PT. 1
INPUT	OUTPUT UNITS: (VDC)

% OF RANGE	UNITS INWC	DESIRED	REQUIRED	AS FOUND	AS LEFT
0	0.0	1.000	0.980 to 1.020		
25	115.2	2.000	1.980 to 2.020		
50	230.4	3.000	2.980 to 3.020		
75	345.7	4.000	3.980 to 4.020		
100	460.9	5.000	4.980 to 5.020		
75	345.7	4.000	3.980 to 4.020		
50	230.4	3.000	2.980 to 3.020		
25	115.2	2.000	1.980 to 2.020		
0	0.0	1.000	0.980 to 1.020		

(Reproduced from McGuire Station procedure.)

Although most plants perform some sort of response time measurements during refueling outages, the practice is not uniform throughout the industry with respect to the number of sensors or channels that are tested and the frequency of tests. Some plants test all their sensors, and others test one-half or one-fourth of their sensors every refueling outage. There are also plants where no sensor response time testing is performed.

A review of the current nuclear industry practice performed in this project shows that most PWRs perform response time testing on both their safety-related RTDs and pressure transmitters, but BWRs only test their pressure transmitters. Thermocouples and neutron detectors are rarely tested for response time.

There is also a wide variation throughout the nuclear industry with respect to response time testing of instrument channels. Basically, the practice varies from no testing to testing all channels at every refueling outage, depending on the plant's safety design and technical specifications.

Typical methods for response time testing of instrument channels, RTDs, and pressure transmitters are described below.

Response Time Testing of Instrument Channels

For response time testing of an instrument channel, a step or ramp test signal is applied at the channel input (simulating a sensor input) while the channel output is recorded. The resulting transient is used to measure the time constant of the channel (for a step input) or the ramp time delay of the channel (for a ramp input). A step signal in the form of a step change in resistance is used for testing the RTD channels and a ramp signal in the form of a ramp voltage or current is used to emulate pressure sensors.

The sensors are usually tested independently of the rest of the instrument channel using a variety of techniques including on-line methods. The on-line methods are advantageous because they can be performed from the control room area in the instrument loops while the plant is operating and can give the response time of the combined sensor and instrument channel in a single test. One of the methods, noise analysis, does not require the channel to be placed in "Test" or by-passed during the response time measurements. It is a completely passive technique and works on a majority of sensors including RTDs, thermocouples, pressure transmitters, and

neutron detectors. The method has been validated for quantitative response time measurements on pressure transmitters, but its use for other sensors is limited to degradation monitoring rather than absolute measurements.

It should be pointed out that response time testing activities in nuclear power plants are not limited to process instrumentation channels or sensors. Other equipment such as valves, breakers, and pumps are also tested periodically to verify proper response time.

Response Time Testing of RTDs

Two methods are available for quantitative response time testing of RTDs. These are referred to as the plunge test and the Loop Current Step Response (LCSR) test. The plunge test is used for laboratory measurement of response time, and the LCSR test is used for in-situ measurement of response time. The plunge test involves a sudden immersion of the RTD from room temperature air into warm water at a flow rate of 3 feet per second. This is the standard method for response time testing of an RTD, but it is only useful for comparative evaluation and selection of RTDs as opposed to measurement of true response time under service conditions. More specifically, the response time results from a plunge test have very little bearing on the in-service response time of the RTD when it is installed in the plant. This is because the response time of an RTD is strongly dependent on the process operating temperature and flow rate. For thermowell-mounted RTDs, in addition to process conditions, the RTD installation and seating in the thermowell is very important. For these reasons, the in-service response time of RTDs can only be identified by in-situ testing at or near normal operating conditions using the LCSR method. This method is based on applying a step change in current to the RTD's extension leads which terminate in the instrument cabinets in the control room area. The current produces an internal heating transient in the RTD that can be analyzed to give the response time. The analysis is based on a heat transfer model for the RTD and a mathematical fitting of the LCSR data to the model. The LCSR method has been approved by the NRC for in-situ response time testing of RTDs in nuclear power plants.

Response Time Testing of Pressure Transmitters

Three methods are available for response time testing of pressure transmitters. These are referred to as ramp test, noise analysis, and the power interrupt (PI) test. The noise analysis and PI tests are used for on-line measurement of response time and the ramp test is used for direct measurements during refueling outages. Unlike RTDs, the response time of pressure transmitters does not depend on process conditions, and the plants have the choice of the ramp method or an in-situ method such as the noise analysis or the PI test. The general practice is to use an in-situ method unless the transmitter cannot be tested by the in-situ method, in which case, the ramp method is used. For example, containment pressure transmitters, cannot be tested on-line with the noise analysis method due to a lack of adequate process fluctuations. These transmitters must therefore be tested by the ramp method, unless they are force-balance transmitters testable by the PI method as described below.

The ramp test uses a hydraulic signal generator to produce a test signal in the form of a pressure ramp. Step pressure signals are also used, but ramp signals are more prevalent in nuclear power plants. This is because the design basis accident analyses in nuclear power plants usually assume pressure transients which approximate a ramp. The ramp test signal is applied to the transmitter under test, and simultaneously to a high-speed reference transmitter. The delay between the output of the two transmitters is the response time of the transmitter under test.

The noise analysis method is based on processing the natural fluctuations (noise) that exist at the output of pressure transmitters when the plant is operating. The noise is recorded from the control room area for a short period of time and analyzed to give the response time of the transmitter. A second method, the PI test, is also available, but this method is applicable only to force-balance pressure transmitters. The PI test is

performed by turning the supply power to the transmitter off for a few seconds, and then on. When the power is turned on, a transient is produced at the output of the sensor that can be analyzed to give its response time.

The validity and accuracy of the noise analysis and PI tests for pressure transmitters have been successfully verified by laboratory measurements performed by AMS under previous research projects for the NRC.[10]

3.5 Acceptance Criteria

The acceptance criteria for the calibration, surveillance, and response time tests are usually given in the plant's technical specifications in terms of a tolerance for the calibration and surveillance tests, and a response time or time constant limit for the response time measurements. In most plants, the tests are performed according to procedures with more stringent acceptance criteria than the plant's technical specifications. Referred to as administrative limits, the more stringent criteria are often imposed by the plant's own management to provide added conservatism.

If a sensor or a channel fails to meet the acceptance criteria for the surveillance tests or the full-channel calibration, adjustments are made to bring the instrument back to tolerance. However, if a component of the instrument channel or a sensor fails the response time tests, the component or the sensor may have to be replaced because there are usually no simple adjustments that can be made to restore the response time, except for such components as low pass filters (lag cards) and certain pressure transmitters with adjustable response times.

4. Smart Sensors

Smart sensors have been growing in popularity ever since they were introduced to the market a decade ago. They are used in a variety of industrial applications and have found their way into the nuclear power industry. Smart sensors are candidates for use in advanced nuclear power plants in addition to conventional plants, and it is therefore important to provide the NRC with an assessment of their capabilities.

As part of the Phase I effort, several smart temperature and pressure sensors were acquired for the project and installed in a laboratory test loop to be evaluated in Phase II. Figure 4.1 shows a photograph of one of the smart sensors being evaluated. This is a Rosemount smart pressure transmitter Model 3051C which has a sensing element with an oil-filled capacitance cell similar to the sensing element of conventional Rosemount pressure transmitters. The output of this sensor is a 4-20 ma current signal proportional to the applied pressure. A digital communication unit that is normally supplied with smart sensors is also shown in Figure 4.1. This is a two-way communication unit with a keypad and display that can be used to read the process parameter on its display or change the sensor configuration from a remote location; as far away as a mile from the sensor. The communication unit is connected in the transmitter current loop as shown in Figure 4.2. The same two wires that provide operating power to the sensor and carry the sensor signal are used to communicate with the sensor. The communication signal is a high frequency AC voltage and its DC average is zero. Therefore, it has no effect on the output of the sensor.

In Figure 4.3 the smart pressure transmitter is compared with its conventional counterpart to show that the physical configuration of smart sensors is somewhat similar but a little smaller than conventional sensors. This is followed by Figure 4.4 comparing a smart pressure transmitter with a smart temperature sensor. The smart temperature sensor is a Rosemount Model 3044C, which can be used with an RTD or a thermocouple. The same communication unit can be used for either pressure or temperature sensors.

Due to the short time that has passed since the smart sensors were acquired for the project, there is no substantial data or conclusions to report about smart sensors other than to state, based on a qualitative evaluation which we have conducted on several Rosemount temperature and pressure transmitters, that smart sensors are easier to use and easier to calibrate than conventional sensors, and they readily meet their specifications as stated by the manufacturer. The smart sensors will be tested thoroughly in the Phase II project, and the results will be presented in the Phase II report.

A smart sensor is essentially a conventional sensor with a built-in microprocessor including memory. The memory is used to store information about the sensor and its calibration. For example, the sensor tag number, serial number, model number, date of last calibration, and other pertinent information about the sensor can be stored within the sensor itself and retrieved at any time through the digital communication unit. More importantly, the sensor calibration constants (e.g., zero, span and range) can be stored in the memory and readily changed to new values when the sensor is recalibrated or reconfigured.

The calibration of a smart sensor is simple in that it does not require manual interaction with the sensor internals. For example, to calibrate a smart pressure transmitter, a pressure corresponding to the sensor's zero is first applied to the sensor and a button is pushed to set the "Zero." Then the full pressure is applied and another button is pushed to set the "Span." These two steps are all that is needed to calibrate a smart pressure sensor as opposed to several steps that must be followed to calibrate a conventional sensor. The zero and span buttons in most smart sensors are accessible from outside the sensor housing, or they can be changed electronically using the sensor communication unit. Therefore, unlike most conventional sensors, there is usually no need in a smart sensor to open the sensor housing for any calibration work. This helps prevent dust, dirt and moisture accumulation on the sensor electronics; a prevalent problem in most conventional sensors. Furthermore, in conventional sensors, to prevent moisture ingress, the housing gaskets or O-rings must be replaced every time the sensor is opened; a time consuming step not necessary in most smart sensors.

A common misconception about smart sensors is in regard to their self calibration or remote calibration capabilities. Even though the terms "self calibration" and "remote calibration" are used by some manufacturers, a smart sensor still requires a known test signal that must be physically applied to the sensor to calibrate it. As such, there is no way to remotely calibrate a smart sensor as installed in a process, other

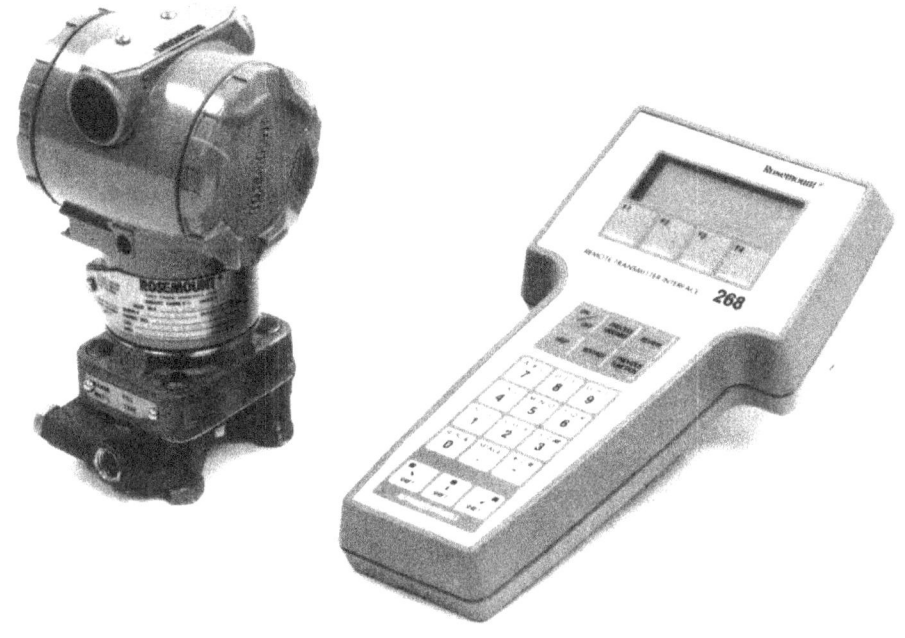

Figure 4.1   Smart pressure transmitter and communication unit

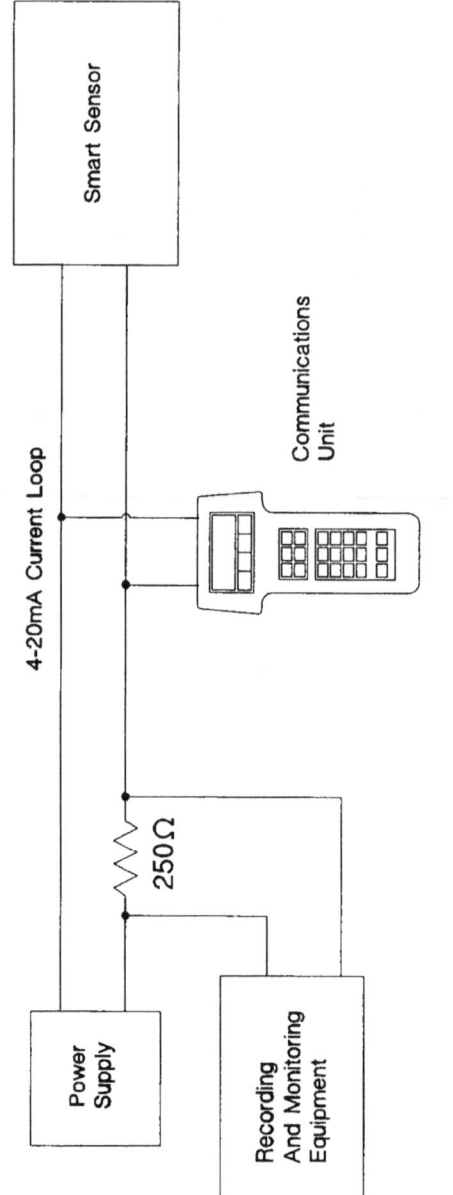

Figure 4.2   Circuit arrangement of a smart sensor

Smart   Conventional

Figure 4.3   Photograph of a smart and a conventional pressure
             transmitter from Rosemount

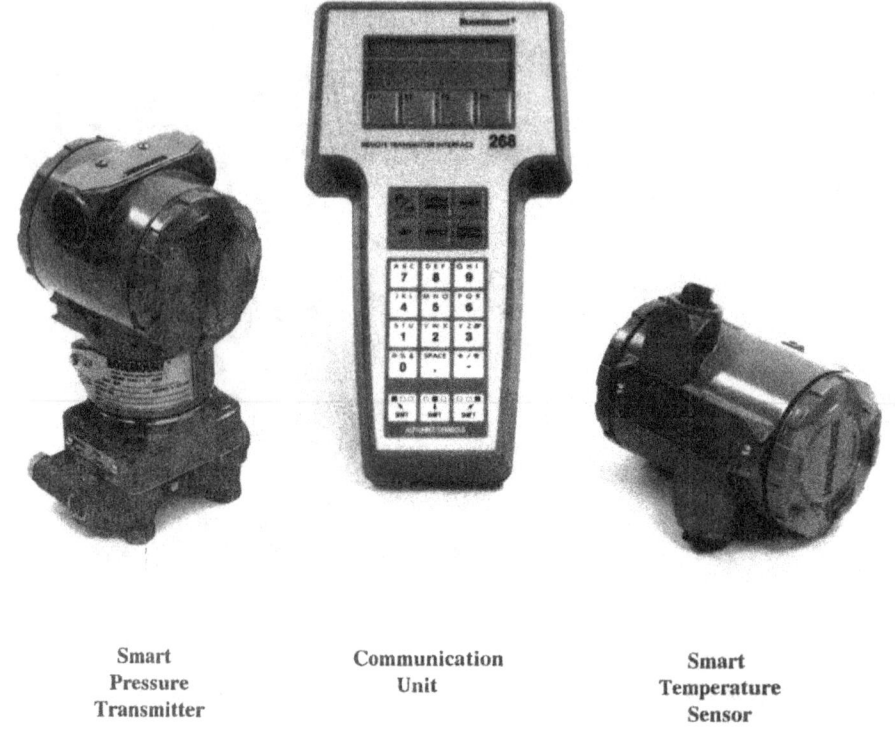

Smart	Communication	Smart
Pressure	Unit	Temperature
Transmitter		Sensor

Figure 4.4   Smart temperature and pressure sensors and communication unit from Rosemount

than to re-range or reconfigure the existing calibration of the sensor.

Figure 4.5 shows a block diagram of a smart sensor. The analog-to-digital converter (A/D) shown in the figure is used to digitize the sensor signal, which is then stored in the memory where it is conditioned as necessary (e.g., linearized, filtered, etc.) or displayed on a digital indicator. The digital-to-analog converter (D/A) is used to convert the digital data (after it has been conditioned) back to analog form, if needed, to be used with existing analog instrumentation in the rest of the instrument channel. A significant advantage of smart sensors is that the sensor signal is digitized in the field before it is sent out to the control room or other areas of the plant. This can provide immunity to electrical noise and other interferences inherent in operating processes if appropriate digital receivers are used. Other important features of the smart sensors are summarized below.

Temperature Compensation

Variations in temperature around a sensor can cause variations in the values of the sensor's electronic components and affect the sensor output. In smart sensors, the ambient temperature variations are measured with a built-in temperature sensor and accounted for in providing the sensor output. This can help reduce temperature errors by nearly an order of magnitude. Temperature compensation is also available in most conventional sensors, but the accuracy of the temperature compensation in smart sensors is by far greater, due to the accuracy of state-of-the-art digital circuitry in the sensor.

Self Diagnostics and Error Reporting

Smart sensors are often programmed to continuously check for gross malfunctions in the sensor such as open circuits and insulation resistance failures, power supply or reference voltage problems, A/D and D/A problems, and microprocessor problems. The sensor can warn the user of a problem or provide error messages. These diagnostic capabilities are particularly useful in isolating hardware problems to a replaceable electronic module or component.

Note that most of the diagnostic capabilities of smart sensors are for their own electronics and digital circuitry,

and there are no provisions to check for the calibration stability and response time characteristics of the sensor. The capability of a sensor to check for its own calibration drift and response time degradation requires sophisticated computing algorithms that cannot be installed on a microprocessor within a smart sensor. Hence, these capabilities can only be added to the existing conventional and smart sensors through the use of an external computer which implements smart sensor technologies. This subject is discussed in detail in the next chapter of this report.

Linearized Output

Once the analog signal from the sensing element is digitized, it can be fit to a linearizing equation stored in the sensor memory. The results can be provided in direct digital form or converted to an analog signal via the built in D/A.

Data Logger

Most smart sensors can store the measurements made for a specified period of time and display them on demand.

Damping

To reduce output noise, smart sensors are often equipped with analog or digital filtering (damping) capability with adjustable response times in the range of 0 to 50 seconds.

Interference and Overvoltage Protection

Smart sensors usually have protective filters to null radio frequency interferences (RFI) and provide immunity to voltage surges (in the case of lightning, for example). They are also equipped with reverse polarity protection.

Tamper Proof

Lockout provisions are often provided in smart sensors to prevent unauthorized tampering resulting in changes to the sensor calibration or range and other characteristics. No pots or knobs are easily accessible on smart sensors thus assuring the sensor configuration control.

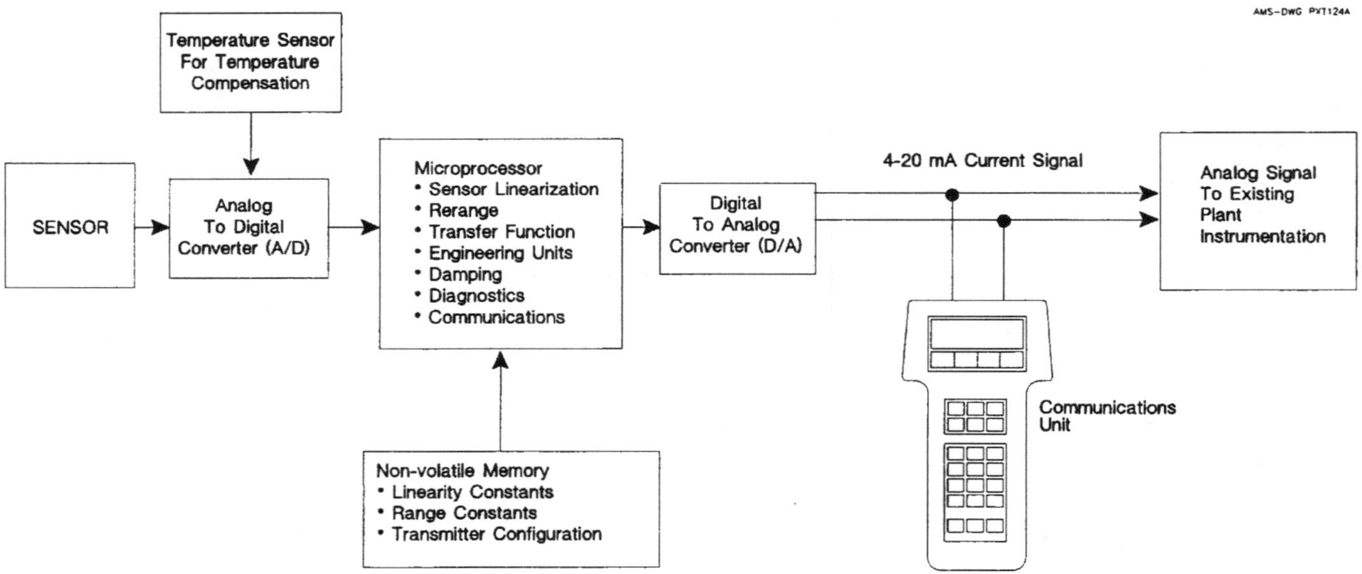

Figure 4.5 Typical components of a smart sensor

5. Smart Sensor Technologies

The existing sensors and instrument channels in nuclear power plants can be adapted to smart sensors using an on-line monitoring system that can sample and analyze the DC and AC output of the instruments while the plant is operating. In addition to providing most of the capabilities that exist in smart sensors, the on-line monitoring system can provide two important features that are not available in smart sensors. These features are:

1. The ability of the instrument to test for its own calibration drift.

2. The ability of the instrument to test for its own response time degradation.

The on-line monitoring system will contain software packages that can analyze the sensor's output to diagnose the health, integrity, and performance characteristics of the sensor and the associated equipment in the signal path from the sensor input to the instrument channel output. These software packages, the algorithms on which they are based, and the associated data acquisition and data processing equipment are referred to here as "smart sensor technologies."

The algorithms which make up the smart sensor technologies include data qualification routines, like signal comparison, analytical redundancy, consistency checking by parity space, and response time degradation monitoring. These algorithms are sometimes referred to as "signal validation" algorithms. Most of the algorithms provide a numerical value (index) that is related to the characteristics of the sensor, instrument channel, or signal being monitored. These indices are calculated once every so often over of a long period of time and plotted as a function of time and reviewed to determine any significant deviations from a normal or desired behavior. Currently, the plots are examined manually. This process will be automated using pattern recognition and neural network techniques to identify the signals which deviate from the expected norms. Eventually all the techniques mentioned here will be integrated into an expert system for on-line testing of calibration drift and response time degradation of process instrumentation channels in nuclear power plants. The expert system will provide a conventional sensor with the intelligence that it needs to function as a truly smart sensor with all the features that are needed in nuclear power plants. The conceptual design of a truly smart sensor is shown in Figure 5.1. The techniques that may be used in such a sensor are summarized in Table 5.1.

An important disadvantage of commercially available smart sensors for nuclear power plants is their susceptibility to radiation damage. The integrated circuits in smart sensors must be radiation hardened if they are to be used in a significant radiation environment. The radiation hardening is an expensive process, and this and other factors are probably the reason why smart sensors have not been qualified by the manufacturers or utilities for nuclear safety-related services. The smart sensor example shown in Figure 5.1 will not suffer from such disadvantages because its integrated circuits are located within an expert system remote from the sensor.

The algorithms and testing techniques mentioned above are described in this chapter. It is important to point out, however, that most of the algorithms discussed here have been developed in the last ten years by various organizations for a variety of applications in nuclear power plants and other industries. In this project, we have brought some of the more common and practical algorithms together and implemented them in an on-line monitoring system to determine their validity, accuracy, and reliability for use in nuclear power plants.

5.1 Data Qualification

Data qualification involves simple computer routines to screen the data and check for undesirable effects such as spikes, sudden jumps, biases, extraneous noise, signal saturation or clipping, and non-symmetric distributions.

For each sensor or instrument channel, a record is established and a file name is designated under which the on-line data is stored. Each data record is divided into a number of segments referred to as data blocks. The sampling rates, the number of data blocks, the number of data points in each block, and other parameters are selected based on the type of signal being monitored, the plant and its operating condition, and whether the system is collecting AC or DC data. Each data record is screened, block-by-block, by calculating, plotting, and examining the mean, variance, skewness, flatness, and other data qualification parameters for each block. These parameters are checked for variations from one block to another and examined against historical values in databases to be built into the on-line monitoring system.

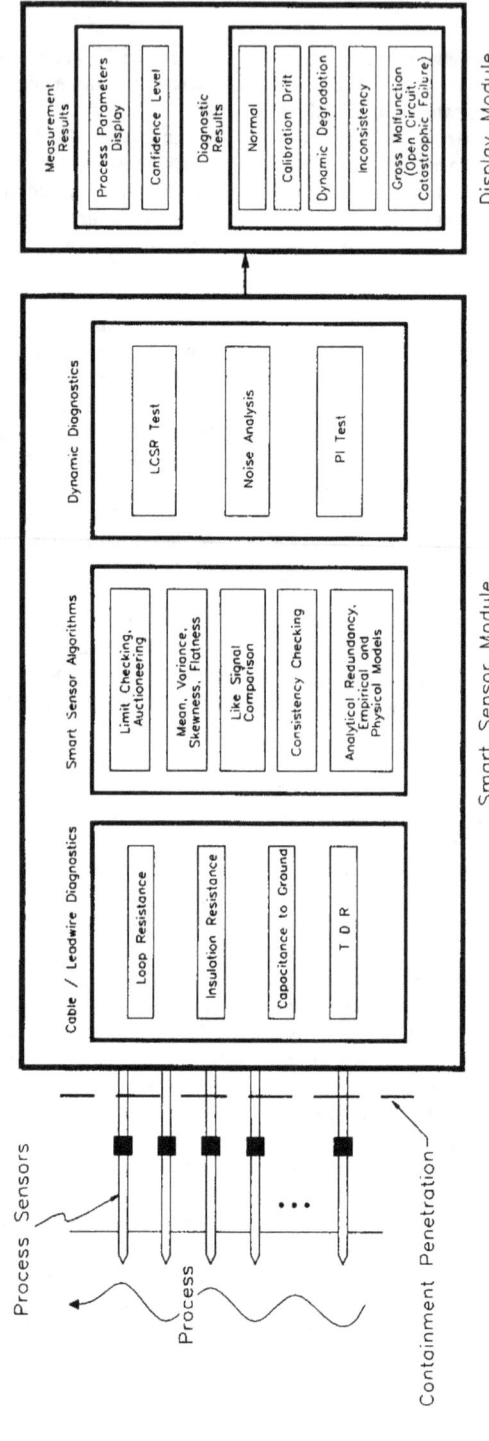

Figure 5.1 Conceptual design of a truly smart sensor system

Table 5.1

Diagnostic Capabilities of a Truly Smart Sensor

Diagnostics Measurement	Application
Loop Resistance	Circuit continuity test
Insulation Resistance	Test of insulation breakdown and moisture ingress
Capacitance to Ground	Test of insulation breakdown and moisture ingress
Time Domain Reflectometry (TDR)	Test of cables and connectors
Limit Checking and Auctioneering	Screening test of DC signals
Mean, Variance, Skewness, and Flatness	Test the quality of AC and DC signals
Like Signal Comparison	On-line testing of calibration of redundant signals
Consistency Checking	Test of consistency between a group of signals
Analytical Redundancy	Adds to hardware redundancy and helps check for common mode problems
Loop Current Step Response (LCSR) Test	Quantitative measurement of response time of RTDs and thermocouples
Noise Analysis	• Quantitative measurement of response time of pressure transmitters • Qualitative testing for response time degradation of temperature, neutron, and other sensors and instrument channels
Power Interrupt Test	Quantitative measurement of response time of force-balance pressure transmitters

The following equations are for a discrete data record that is represented by an array of discrete signal values of the form $x(1)$, $x(2)$, $x(3)$, ..., $x(N)$.

Mean

The mean (m) or average of a data block is calculated by adding all the data points in the block and dividing the sum by the number of points in the block (N).

$$m = \frac{1}{N} \sum_{k=1}^{N} x(k) \qquad (5.1)$$

Examples of problems that can be identified by monitoring the mean value are sudden loss of signal, signal saturation, etc.

Variance

Variance (σ^2) is a measure of the dispersion of the signal about its mean. The variance for each block is determined by calculating the block mean and subtracting each data point from the mean. The result is then squared, summed, and divided by the number of data points in the block.

$$\sigma^2 = \frac{1}{N} \sum_{k=1}^{N} [x(k) - m]^2 \qquad (5.2)$$

A related term to variance is root mean square or RMS. RMS is also called standard deviation (σ). It is the square root of variance.

$$RMS = \sigma = \sqrt{\frac{\sum_{k=1}^{N} [x(k)) - m]^2}{N}} \qquad (5.3)$$

An example of a problem that can be identified by monitoring the variance is abnormally large or small signal fluctuations.

Skewness

Skewness is a measure of signal symmetry about its mean value. For a completely symmetrical signal, the skewness value as calculated by the following equation must be zero.

$$\mu_3 = \frac{1}{N} \sum_{k=1}^{N} (x(k) - m)^3 / \sigma^3 \qquad (5.4)$$

Experience has shown that signals from properly operating instruments in nuclear power plants usually have skewness values close to zero. The skewness of a signal can be seen in a plot of the amplitude probability density (APD) function for the signal. The APD plot is similar to a histogram (Figure 5.2). A histogram is typically plotted as follows: 1) on the y-axis, the number of times the same outcome is attained, and 2) on the x-axis, the range of possible outcomes. In plotting the APD of a signal, the signal variations in amplitude are divided into a number of discrete values and the probability of occurrence of each amplitude is plotted as a function of the amplitude. A properly distributed signal will have a bell-shaped APD as shown in Figure 5.2. In Figure 5.3, we have shown a normal APD and a skewed APD.

The bell-shaped distribution is represented by the following equation:

$$p(x) = \frac{\exp(-x^2 / 2\sigma^2)}{\sigma\sqrt{(2\pi)}} \qquad (5.5)$$

where $p(x)$ is the probability of occurrence of the signal x and σ is the RMS value of the signal. A bell-shaped curve that follows Equation 5.5 is referred to as a Gaussian distribution. A Gaussian distribution is also called a "normal" distribution. If the skewness is not zero, then the signal is said to be skewed, implying that it is not symmetrical about its mean value.

A departure from the Gaussian distribution can be determined by calculating the skewness of the signal or comparing the APD plot of the data against a Gaussian APD. A departure from the Gaussian distribution is usually indicative of nonlinearity in the signal or sensor from which the signal originates. Examples of problems that may be detected by calculating the skewness or plotting the APD are signal clipping, nonlinearity, one-sided signals, gain problems, etc.

Flatness

Flatness (also called Kurtosis) is calculated by the following equation:

$$\mu_4 = \frac{1}{N} \sum_{k=1}^{N} [x(k) - m]^4 / 3\sigma^4 \qquad (5.6)$$

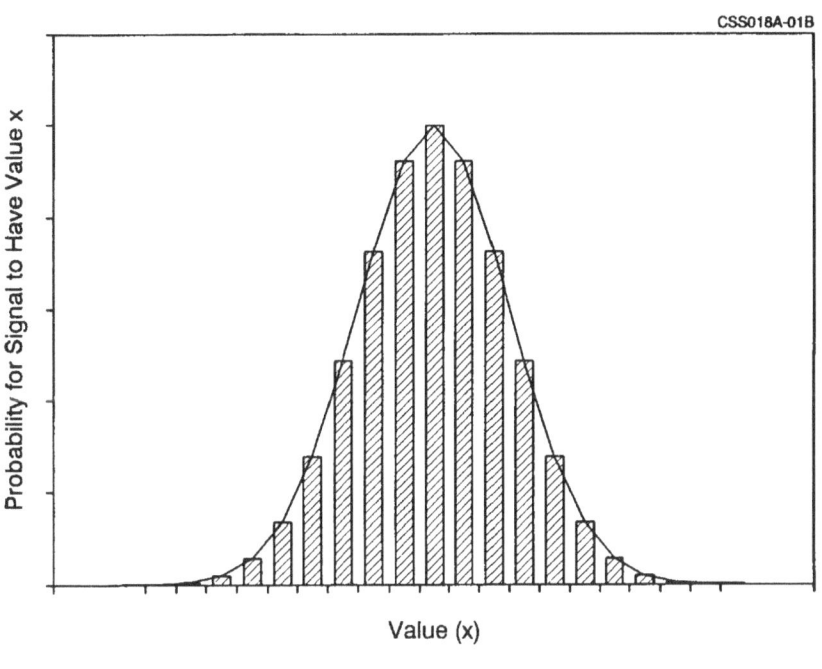

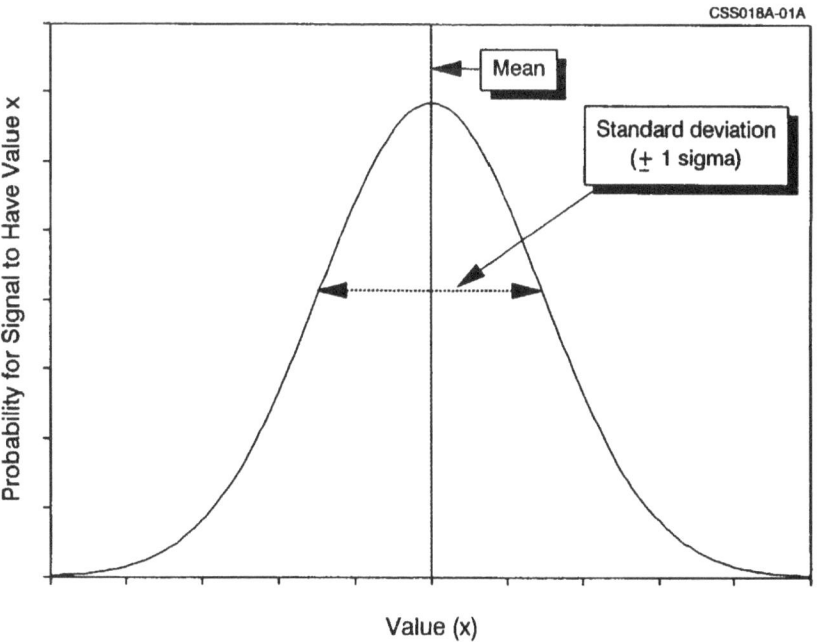

Figure 5.2   Illustration of a histogram (top) and an APD (bottom)

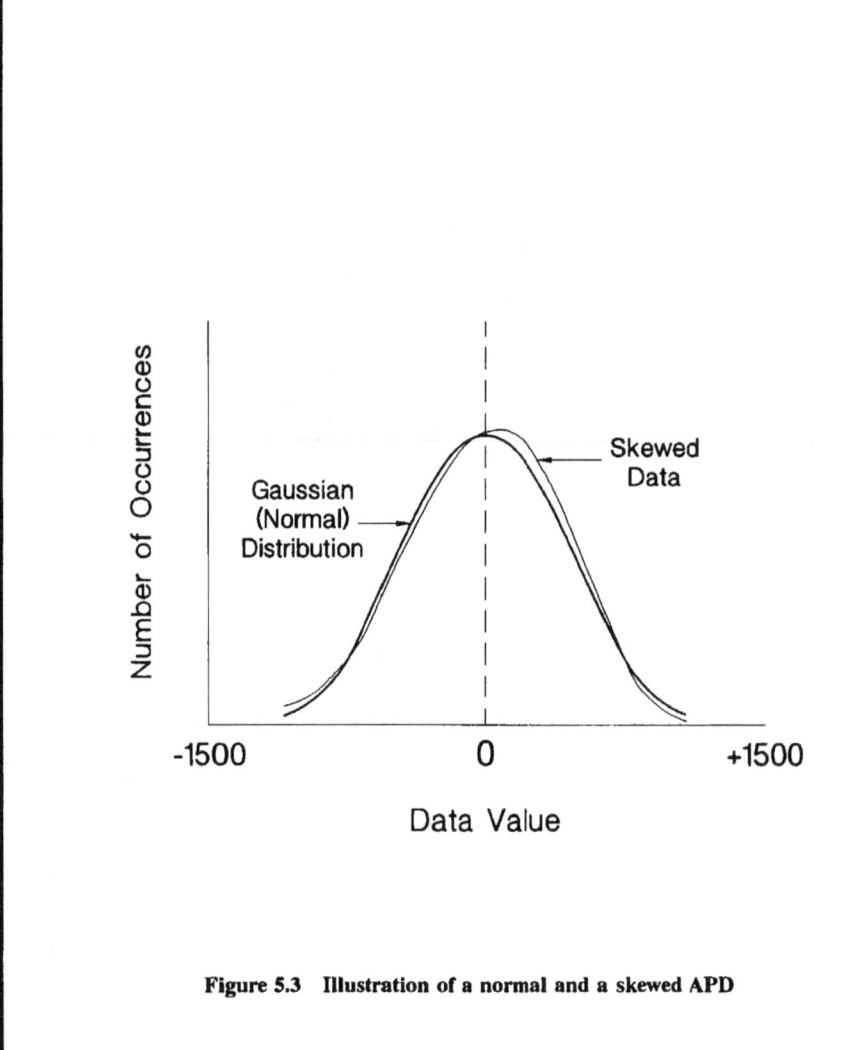

Figure 5.3   Illustration of a normal and a skewed APD

The flatness must have a value close to unity for a properly behaving signal. The flatness is basically a measure of the ratio of small amplitude occurrences to large amplitude occurrences in the signal. Flatness can also be evaluated from the shape of the APD. Figure 5.4 shows the APDs of a slightly and a highly non-flat signal.

Plot of Data Qualification Parameters

The data qualification parameters mentioned above are plotted block-by-block for a normal and an abnormal signal in Figures 5.5 and 5.6, respectively. It is apparent that the normal signal has a relatively constant mean and variance for all the data blocks, a normal APD and a flatness close to unity for all blocks. In contrast, the abnormal data show saturation between blocks 80 and 90 as indicated by a jump in the mean and variance values, a spike in the APD plot and a flatness value much different than unity.

The mean, variance, skewness, and flatness are referred to as statistical moments of the signal and defined based on signals with random fluctuations that follow a Gaussian distribution. These moments are also called 1st, 2nd, 3rd, and 4th moments corresponding to mean, variance, skewness, and kurtosis, respectively. There are higher moments such as 5th, 6th and higher, but no specific name is used for these higher moments. The mean, variance, skewness, and the higher moments are useful mainly for qualification testing of AC signals, but long records of slowly varying DC signals can also be qualified by these moments. In addition to calculating the statistical moments, limit checking and auctioneering approaches are used for DC signals. In this project, we have used a method similar to limit checking for identifying and deleting the discontinuities in the in-plant test data that are encountered when the plant trips during the on-line monitoring process.

5.2 Like Signal Comparison

Like signal comparison is a useful method for detecting calibration problems or drift in redundant sensors or instrument channels. Two or more signals are referred to as "like" signals if they measure the same process parameter. For example, the RTDs that measure the hot leg temperature in the primary coolant system of a PWR are referred to as "like" signals. Two or more signals that are not "like" signals are referred to as diverse, dissimilar, or "unlike" signals.

The procedure for drift monitoring using the like signal comparison involves averaging the redundant signals and determining any significant deviation of each signal from the average of all signals. Straight averaging is used unless the redundant signals are from sources of varying quality or accuracy. For example, if a group of RTDs and thermocouples are tested together, it may be appropriate to weight the RTD signals more than the thermocouple signals because RTDs are generally more accurate than thermocouples. The weighting factor is determined based on: 1) previous experience, 2) general information about the accuracy of instruments, 3) information about the age and accuracy of the sensor calibration, and 4) consistency of the redundant signals, or a combination of these and other factors.

The accuracy and reliability of the like signal comparison analysis depends predominantly on the number and reliability of the redundant signals available. The minimum number of signals necessary is three and the more signals the better. With as few as three signals, it is important to perform pairwise comparisons using a consistency checking method to ensure that the drift in one of the three signals does not invalidate the average value. If it does, the signal must be labeled as an "outlier" and removed from the average. The deviation of the outlier is then identified by determining its difference from the average of the remaining signals. When less than three signals remain after the outlier(s) are removed, the analytical redundancy approach as described below may have to be employed to make up for the lack of physical redundancy.

Figure 5.7 shows five traces for temperature sensors installed in our laboratory test loop to measure the same temperature. These time traces are useful in studying the steady-state behavior of the instruments. Figure 5.8 shows three pressure transmitter signals of which one transmitter was drifting and causing all three transmitters to show a drift. A pairwise comparison of the signals helped identify the drifting transmitter and separate it from the stable transmitters (Figure 5.8., top right).

The like signal comparison technique is widely used in pressurized water reactors (PWRs) for on-line testing of calibration of primary coolant RTDs. The tests are performed remotely from the control room area when the plant is at isothermal conditions. At isothermal conditions, the primary coolant RTD elements in the hot legs and cold legs of the plant are at the same temperature. This provides about 20 to 40 temperature elements that are monitoring the same temperature and can therefore be intercompared to identify the outliers. The test is referred to as the cross calibration test which

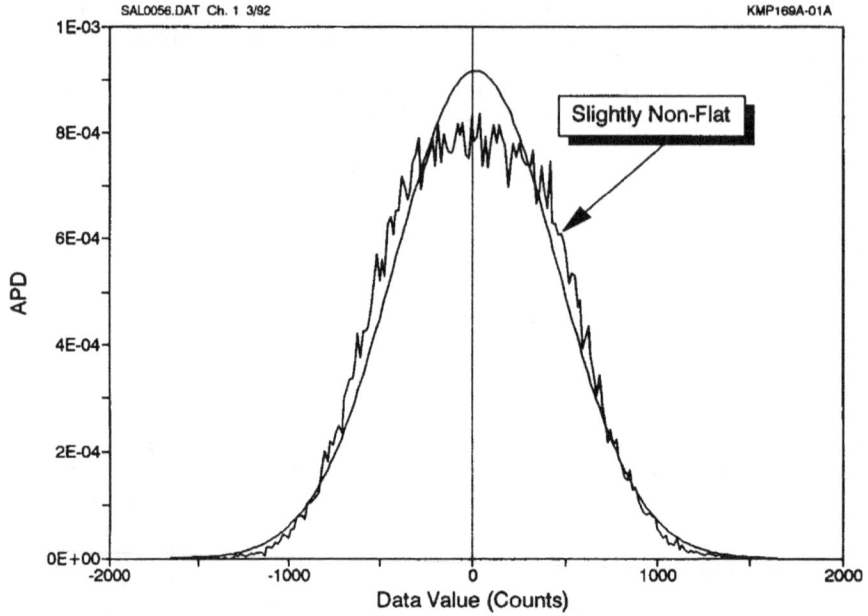

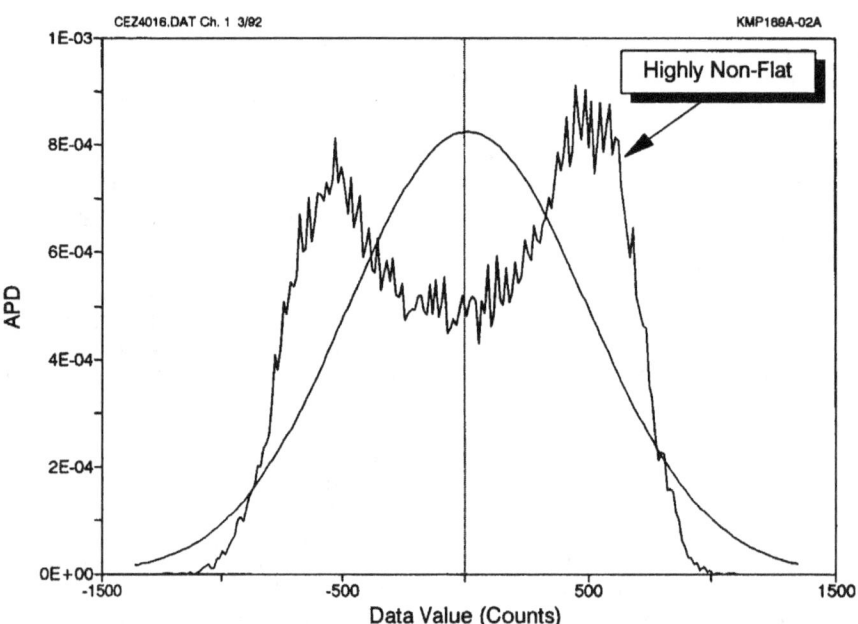

Figure 5.4 Amplitude Probability Density plots for slightly and highly non-flat signals

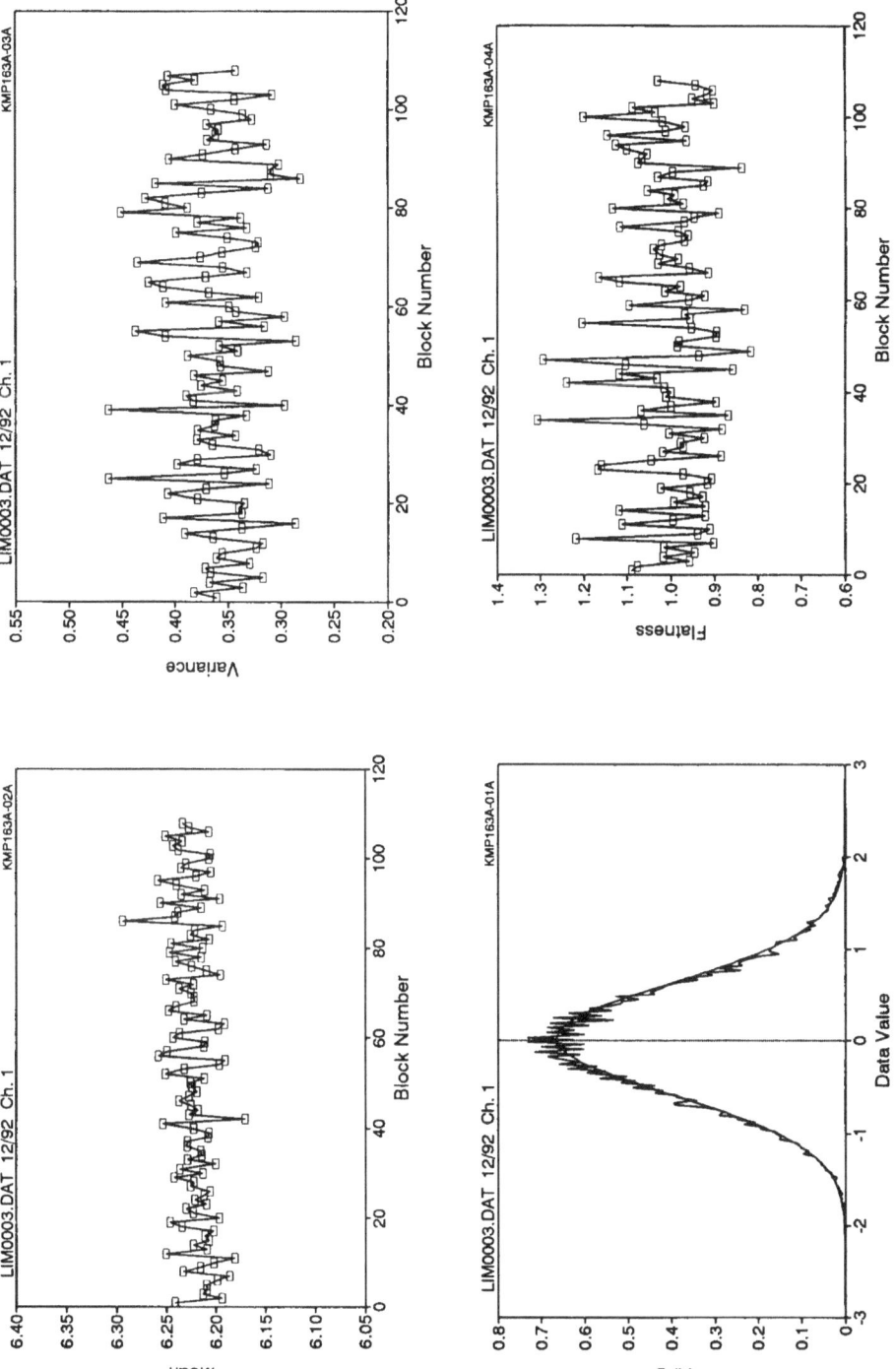

Figure 5.5 Data qualification plots for a normal signal

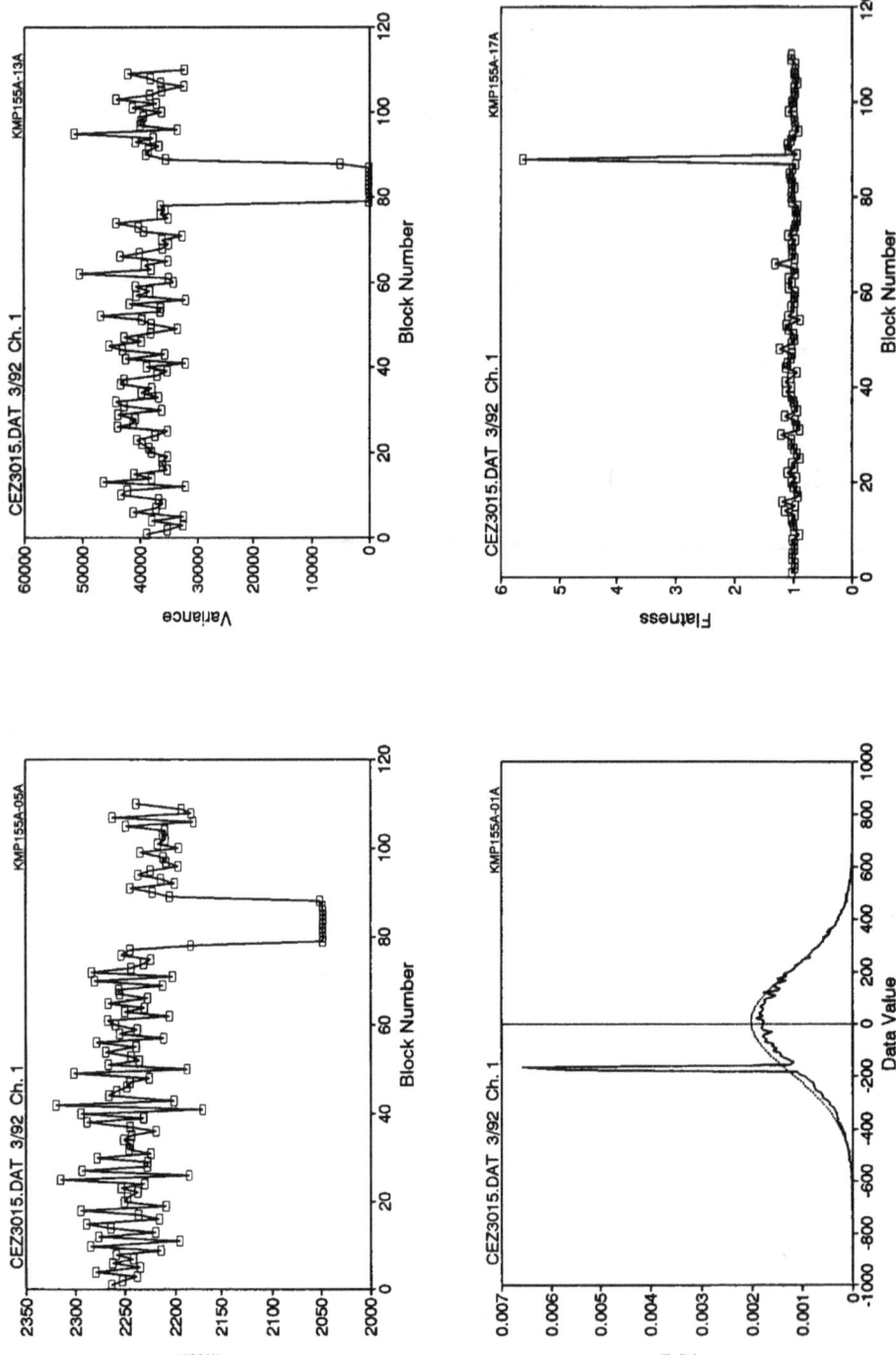

Figure 5.6 Data qualification plots for an abnormal signal

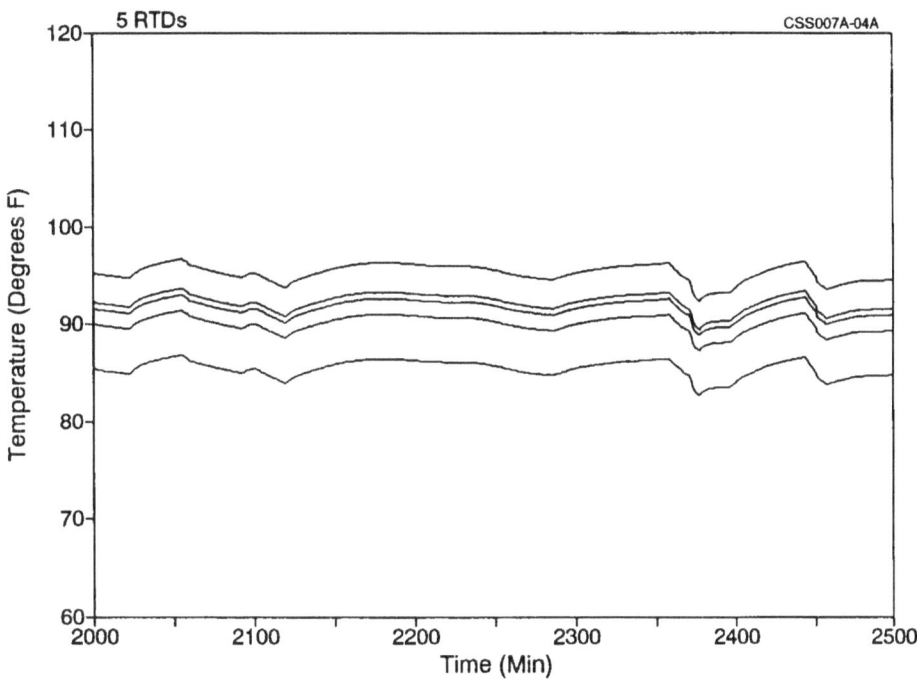

Figure 5.7 Time history plots for five temperature sensors tested in
the laboratory test loop

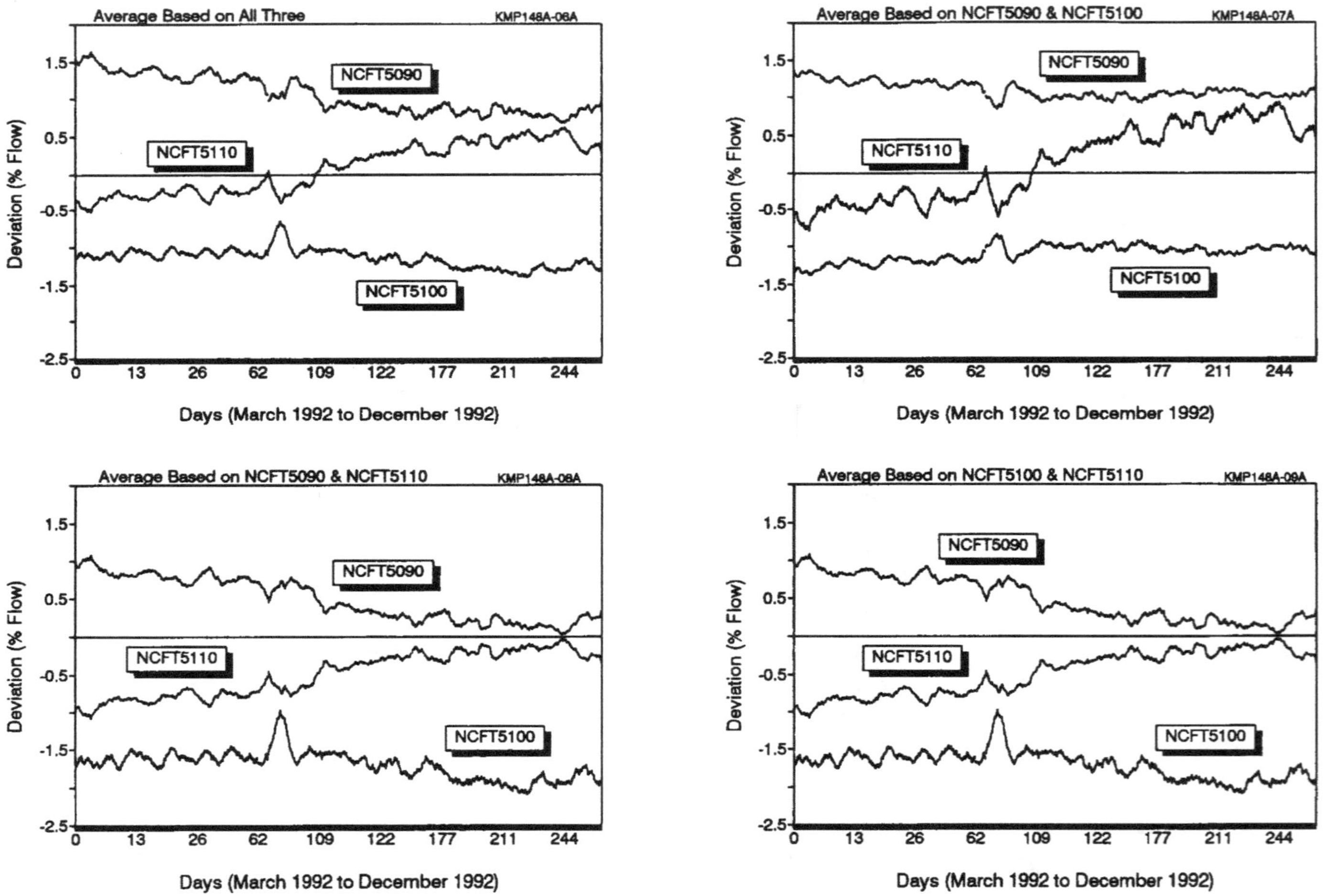

Figure 5.8 Time history plots from in-plant testing of three differential pressure transmitters

is described in detail in Chapter 6. As will be seen in Chapter 6, the redundancy of RTDs for cross calibration is so good that it is not necessary to plot the data over time. A few data scans are usually adequate for a conclusive test.

5.3 Analytical Redundancy

This method is used for obtaining an independent estimate of a process parameter to account for common-cause failure or unidirectional drift. More specifically, if a group of sensors drift upward or downward together and away from the true process variable, the intercomparison technique cannot identify the problem. Therefore, it is important to use an additional measure to obtain an independent estimate of the process parameter and ensure that common mode problems are accounted for. In addition to helping to account for common mode drift, analytical redundancy is a method of generating redundant signals from non-redundant sensors (Figure 5.9).

Analytical redundancy takes advantage of the fact that in a closed loop system such as a nuclear power plant, many of the process parameters are often closely related. For example, an estimate of the reactor coolant flow in a PWR can readily be obtained from measurement of hot leg and cold leg temperatures and the reactor power. This can be done using the $Q = \dot{m} c_p \Delta T$ formula, where Q is the reactor thermal power, ΔT is the temperature drop across the core, c_p is the constant pressure specific heat and $\dot{m}$ is the flow rate of interest. Similar information can be obtained from an empirical correlation of the form:

$$Y = C_o + \sum_{i=1}^{n} C_i \, \Phi_i \, (\underline{x}) \qquad (5.7)$$

where Y is the parameter of interest, $\underline{x} = (x_1, x_2, ..., x_n)$ is the vector of variables that influences Y, $\{\Phi_i, i=1,2, ... \, n\}$ are nonlinear terms, and $\{C_i\}$ are constants. The constants of Equation 5.7 can be identified through a so-called learning process. The learning process is like fitting on-line process data to Equation 5.7 and solving a system of equations that gives the constants of the equation. This will provide the characteristic equation for the process that can be used later to predict the values of the process parameters from on-line measurements of other parameters. The advantage of empirical modeling over physical modeling is that it does not require a knowledge of the system structure and material properties and a disadvantage is that it requires the

model to be validated for every process at several steady-state conditions. The accuracy of the empirical and physical models depends on the validity of the model for the process being analyzed, the process operating conditions, the number of diverse signals being analyzed, and other factors.

Examples of how empirical and physical models are implemented are discussed below.

Implementation of Empirical Models

A number of organizations have worked on the development of algorithms for analytical redundancy. One of the most notable developments in this area has been carried out by the University of Tennessee (UT) under a contract with the U.S. Department of Energy.[11,12] The UT researchers have developed and tested the empirical model of Equation 5.7 with operating data from a four loop commercial PWR. Two examples of UT's results for this PWR are described below.

1. The reactor power was modeled as a function of the hot leg and cold leg temperatures with the following result:

$$y = 0.00254x_2^2 - 64.5x_1 + 0.0568x_1^2 \qquad (5.8)$$
$$- 1.18x_2 + 18178$$

where y = reactor power (%)
          x_1 = cold leg temperature (°F)
          x_2 = hot leg temperature (°F)

Figure 5.10 shows a plot of Equation 5.8 compared with actual measurements of power from 0 to 100%. It is interesting that the same model can reasonably predict the output of the system for a wide operating range from zero power to 100 percent power operation.

2. The pressurizer level was modeled as a function of pressurizer pressure, reactor power, hot leg temperature and cold leg temperature. The prediction model is:

$$y = 0.7365x_4 - 0.0685x_1 - 0.0086x_2 \qquad (5.9)$$
$$+ 0.64x_3 - 723$$

where y = pressurizer level (%)
          x_1 = reactor power (%)
          x_2 = pressurizer pressure (lb/in^2)
          x_3 = cold leg temperature (°F)
          x_4 = hot leg temperature (°F)

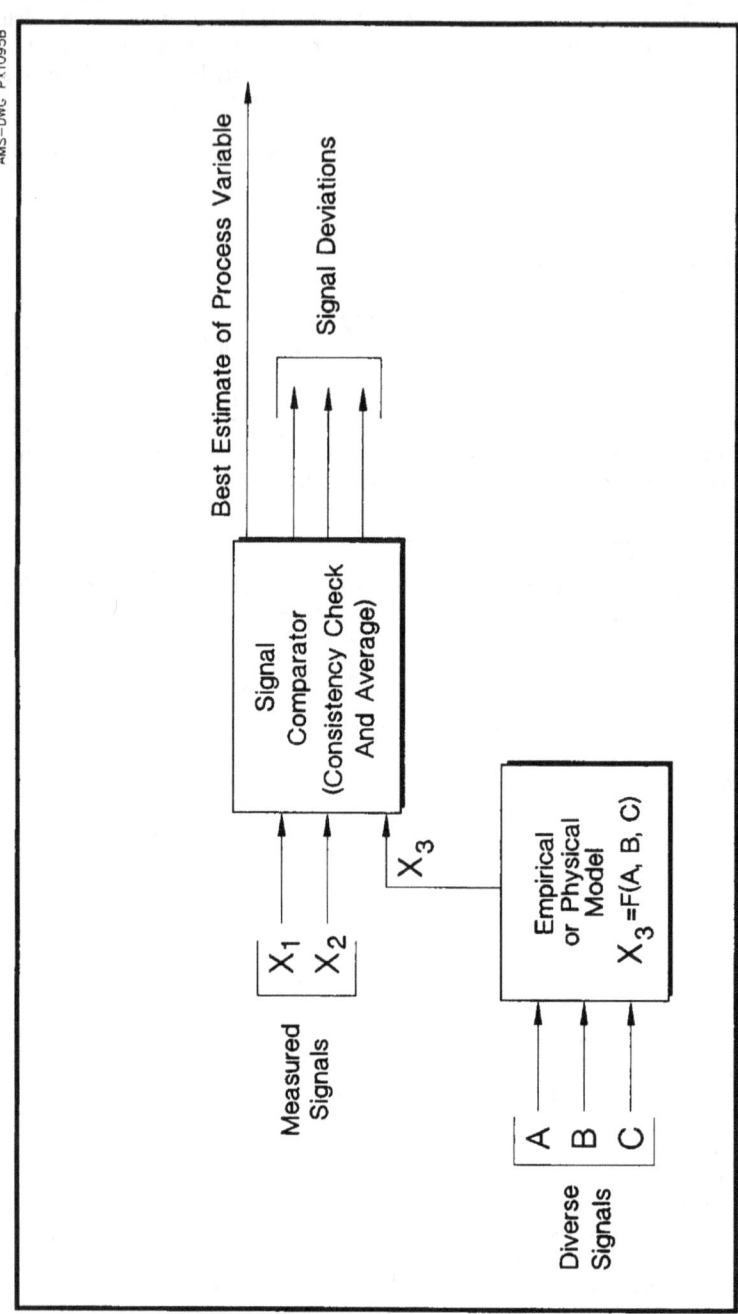

Figure 5.9   Illustration of principle of analytical redundancy

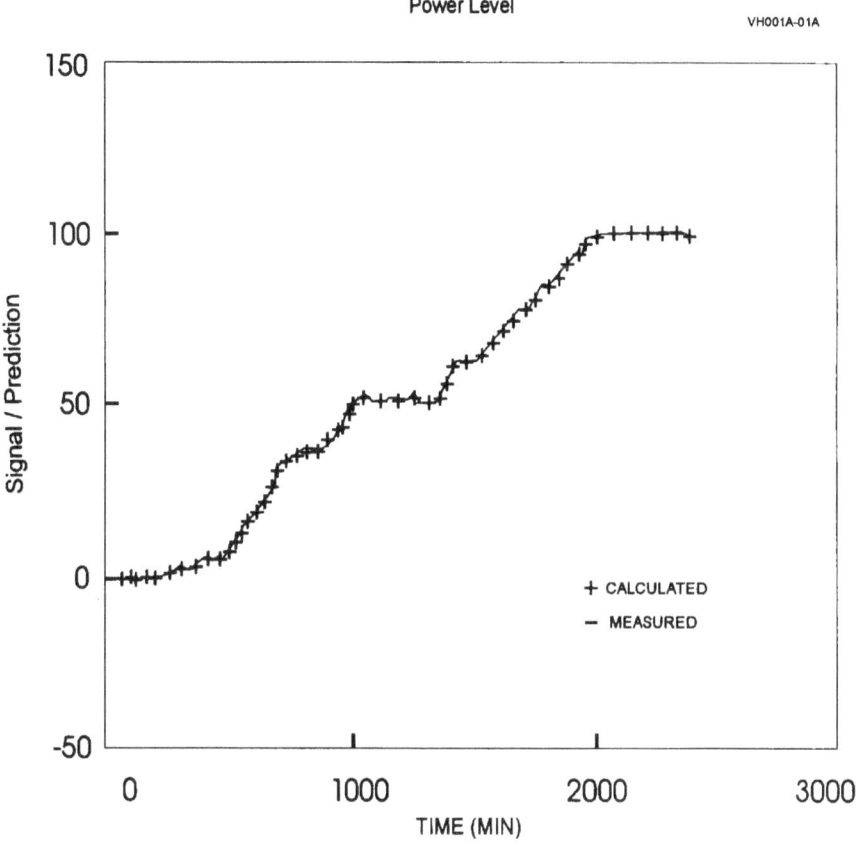

Figure 5.10 Empirical model prediction and actual measurement of reactor power in a PWR

Figure 5.11 illustrates the performance of the model compared with the measurement of pressurizer level. Note that the model can correctly predict the output of the system throughout its operating range.

Implementation of Physical Models

A simple physical model was developed in Phase I and used with laboratory data to predict the temperature of a test loop simulating a PWR. The test loop is shown in Figure 5.12. During normal operation, flow passes through both the high and low velocity legs, while the by-pass portion of the system is isolated. Temperature in the loop is regulated by directing a portion of the flow to a heat exchanger located on the discharge side of the recirculation pump. The heat exchanger is exposed to cool outside air and is effective in removing heat generated in the system by the recirculation pump. During steady-state operation, test loop conditions are as follows:

Operating Temperature:	45°F to 120°F
Operating Pressure:	45 to 200 PSIG
Main Loop Flow Rate:	400 to 450 GPM
Heat Exchanger Flow Rate:	0 to 50 GPM

By the first law of thermodynamics, neglecting any heat transfer from the piping to ambient air, the rate of change in internal energy $(\dot{E})$ of the system (consisting of the water in the entire loop) must equal the sum of the rate of heat removed by the heat exchanger $(\dot{Q}_{HX})$ and the rate of heat added by the recirculation pump $(\dot{Q}_{pump})$.

$$\dot{E}_{net} = \dot{Q}_{HX} + \dot{Q}_{pump} \quad (5.10)$$

Thus,

$$mc\,(T_2 - T_1) = (\dot{m}_{HX}\,c\,\Delta T_{HX} + \dot{Q}_{pump})\,\Delta t \quad (5.11)$$

where m = total mass of water in the loop
   $\dot{m}_{HX}$ = mass flow rate through the heat exchanger
   T_1 = initial temperature
   ΔT_{HX} = temperature change of water across the heat exchanger
   T_2 = final temperature
   c = specific heat capacity of water.
   Δt = increment of time

Solving for T_2:

$$T_2 = T_1 + \left[\frac{\dot{m}_{HX}\,c\,\Delta T_{HX} + \dot{Q}_{pump}}{mc}\right]\Delta t \quad (5.12)$$

The model calculates the final loop temperature (T_2) on a "per minute" basis. Once the final loop temperature is calculated for the first minute, this value is used as the new initial temperature of the test loop (T_1).

The test of the model was performed using heat exchanger flow rates ranging from 0 to 25 GPM and outside air temperatures ranging from approximately 49°F to 72°F. This provided a broad range of conditions to evaluate the performance of the model during steady-state and transient conditions. The results of a test run comparing the model predictions and measured temperatures are shown in Figure 5.13. The differences between the two readings are shown in Figure 5.14. It is apparent that the model accurately predicts the loop temperature within better than $\pm 1°F$. This difference increases slightly as the process temperature increases above 85°F or drops below 70°F when heat transfer to and from the piping to the ambient air becomes more significant.

The simple model and the laboratory tests described here show the feasibility of the physical modeling approach for independent assessment of process parameters. In Phase II, a more sophisticated model that will be suitable for predicting a plant signal behavior will be developed and validated in the test loop and then incorporated in the on-line monitoring system.

5.4 Consistency Checking

As mentioned earlier, if only a few "like" signals are available for intercomparison, a drift in one of the signals will cause the other signals to appear as if they are drifting. To avoid this problem, each signal is individually compared with other signals for consistency. If the difference between any two signals exceeds a threshold, an inconsistency is declared. Furthermore, an inconsistency index is established to count the number of times that a signal has been inconsistent with one or more other signals. The inconsistency index is incremented by one each time an inconsistency is declared. The inconsistency index is then used as a weighting factor in averaging a group of redundant

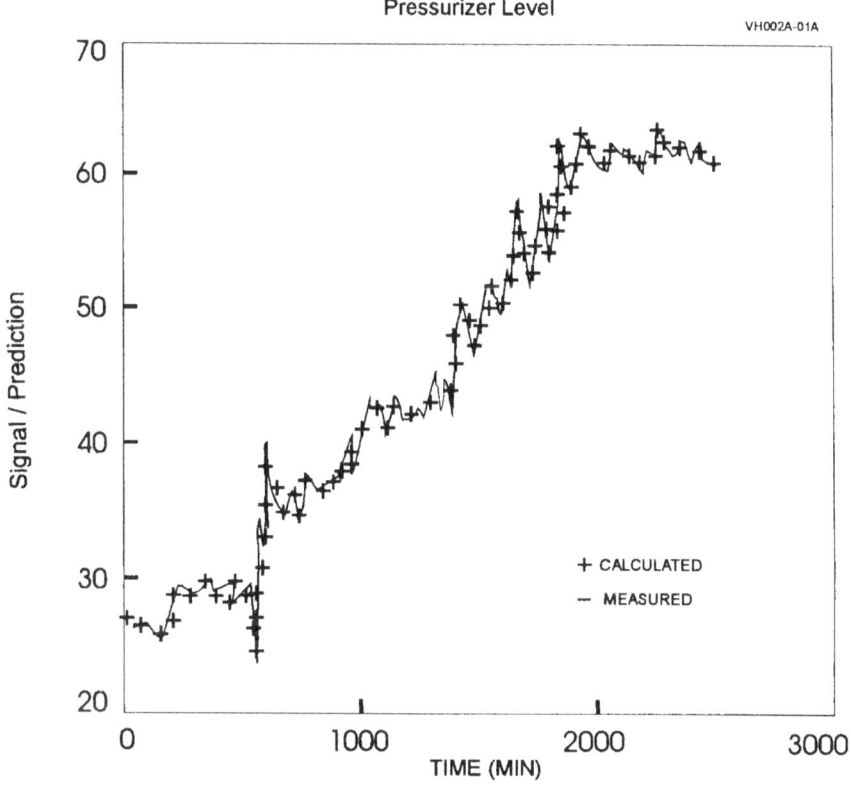

Figure 5.11 Empirical model prediction and actual measurement of pressurizer level in a PWR

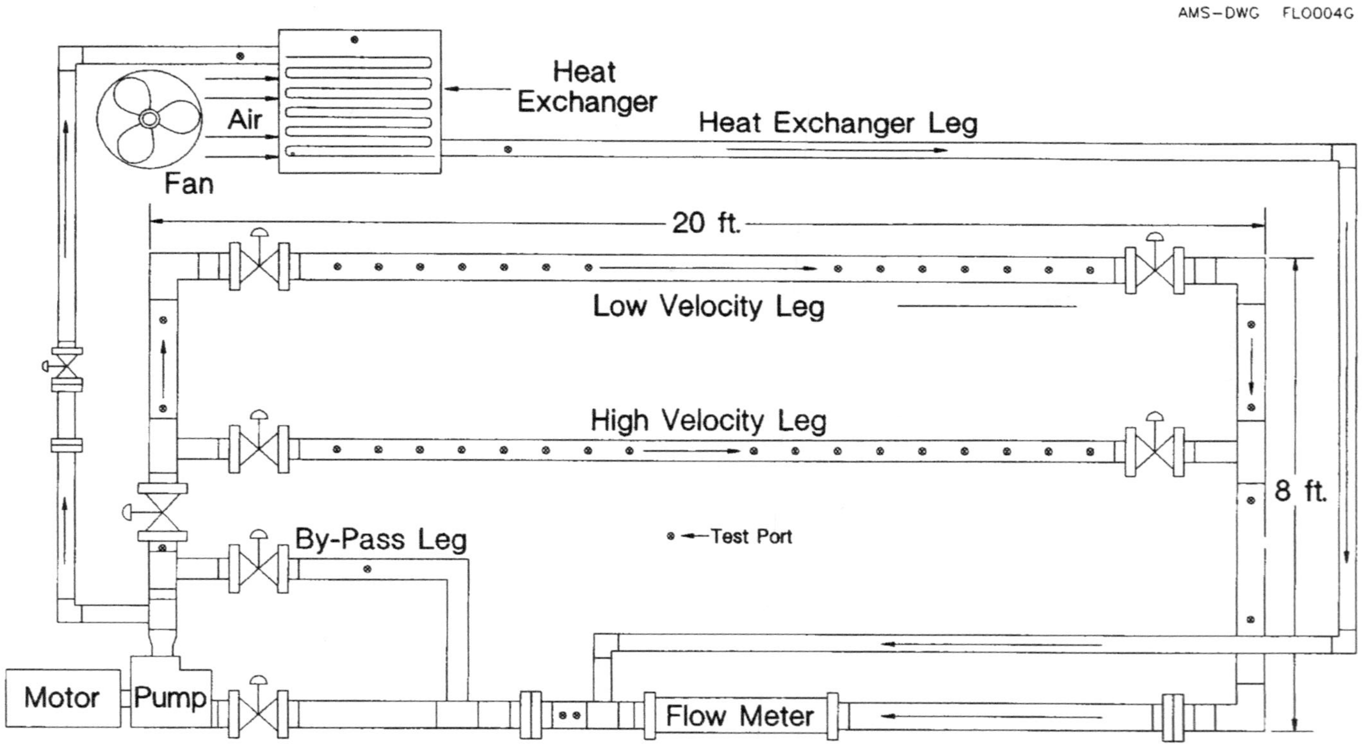

Figure 5.12 Diagram of laboratory test loop

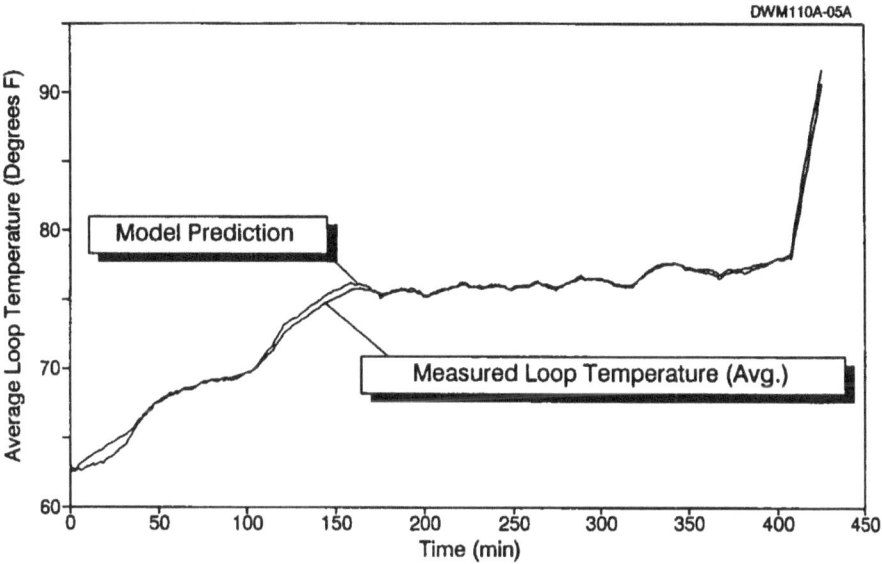

Figure 5.13   Physical model prediction and actual measurement of
temperature in laboratory test loop

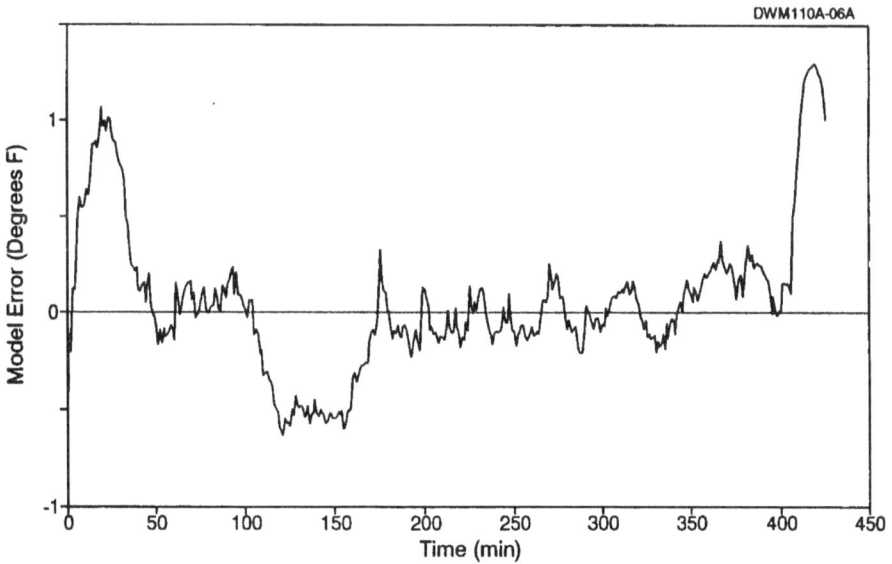

Figure 5.14   Difference between model prediction and actual
measurement of loop temperature

signals to provide a best estimate for the process parameter.

Figure 5.15 shows monitored data for five temperature sensors in our laboratory test loop. One of the sensors (TC-I-6) is inconsistent with the others as shown in the figure. The inconsistency is marked on the figure by placing an " * " on the trace every time the signal is inconsistent with any other signal in the group.

Inconsistency checking is a useful operation regardless of how many like signals are being monitored together. This method, or a variation of it, is referred to in the literature as the parity space technique. However, most parity space algorithms are written in terms of vectors as opposed to scalars, which we have used here. We will explore the potential of using other variations of parity space in Phase II if it is determined that they are more effective than the one we have implemented here.

In performing consistency tests, it is important to account for the tolerance band of the signals being compared with each other. If the two signals are designated as $x(k)$ and $y(k)$ with the tolerance of x being ϵ_1 and tolerance of y being ϵ_2, then the consistency criteria may be written as:

$$|x(k) - y(k)| \geq \epsilon_1 + \epsilon_2 \qquad (5.13)$$

That is, if at any time the difference between the two signals (neglecting the sign) is greater than the sum of the tolerances of the two signals, then the two signals are said to be inconsistent and the inconsistency index for each signal is increased by 1.

It should be pointed out that any group of signals can be very consistent with one another and still provide an inaccurate estimate of the process parameter. For this reason, it is always important to provide an independent estimate of the process parameter using a form of analytical redundancy.

5.5 Uncertainty of Calibration Testing Techniques

The calibration testing method discussed above will only work if the uncertainties of the techniques are less than the deviation that we are trying to resolve. These uncertainties have not been quantified to date. Therefore, there is currently no way to know if on-line testing of calibration of sensors or instrument channels in nuclear power plants can become a reality. What is certain, however, is that gross changes in a sensor

calibration can readily be identified by on-line monitoring even though it is not yet possible to quantify the change with adequate accuracy. This by itself is very useful in determining ahead of time which sensors or instrument channels would need maintenance or replacement. It is also certain that the uncertainties of on-line calibration techniques can be reduced if a combination of techniques are used together. Figure 5.16 shows a block diagram of a system that uses all the techniques discussed in this chapter including the response time degradation monitoring techniques summarized below. The combination of all of these techniques should provide a more reliable estimate of the process parameter than any one technique. If all techniques are applied to each data block for the life of a fuel cycle, it is reasonable to expect that reliable information about the performance of the instrument channel should be attainable.

5.6 Response Time Degradation Monitoring

The AC data from the on-line monitoring system can be analyzed to monitor for changes in response time of sensors or instrument channels. This method assumes that the AC component of process input to the sensor is wideband random noise whose spectral properties remain essentially the same throughout the monitoring process. The analysis can use simple methods such as zero crossing or more sophisticated methods such as univariate Autoregressive (AR) modeling or Fast Fourier Transform (FFT) techniques to produce the Power Spectral Density (PSD) of the AC signals to be monitored for changes. The combination of these methods are referred to as "noise analysis." These methods are described below.

The word "noise" refers to the random fluctuations in a process parameter as opposed to high frequency electrical noise, 60 cycle noise, and other electrical interferences. The frequency range of interest in noise analysis testing of sensors in nuclear power plants is typically 0 to 100 Hz.

Zero Crossing

Zero crossing involves counting the number of times that a noise signal crosses its average value in a given period of time. The average value is usually zero because it is customary to remove the DC component of a signal before it is used for response time monitoring. The DC component is removed by a variety of techniques such as high-pass filtering, adding a DC offset, or measuring the DC value of the signal after it

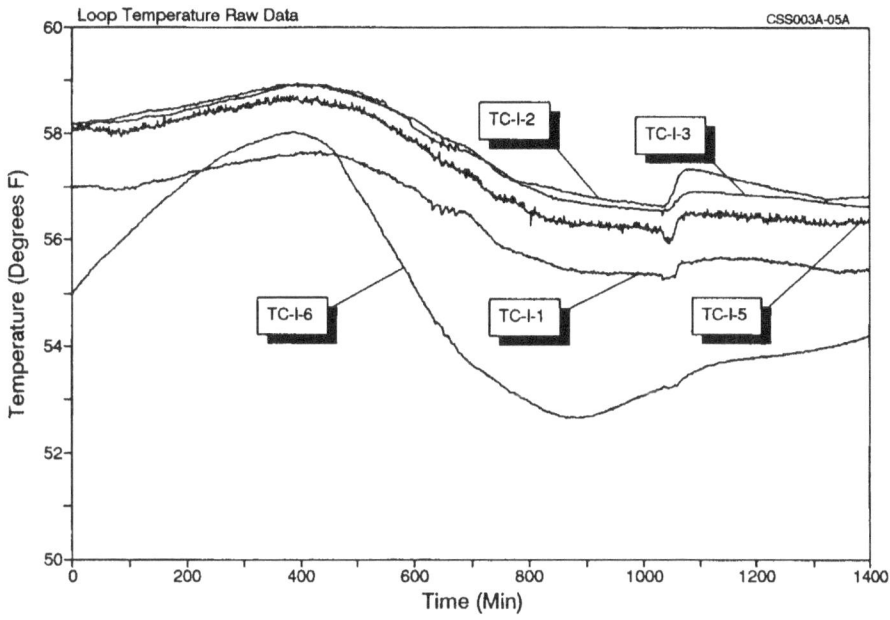

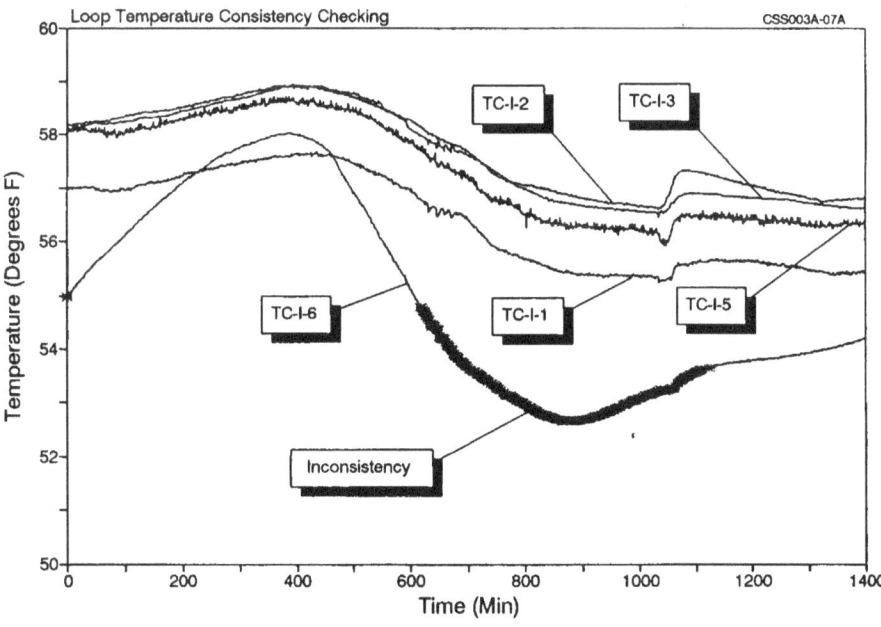

Figure 5.15   Results of consistency checking performed on five temperature signals

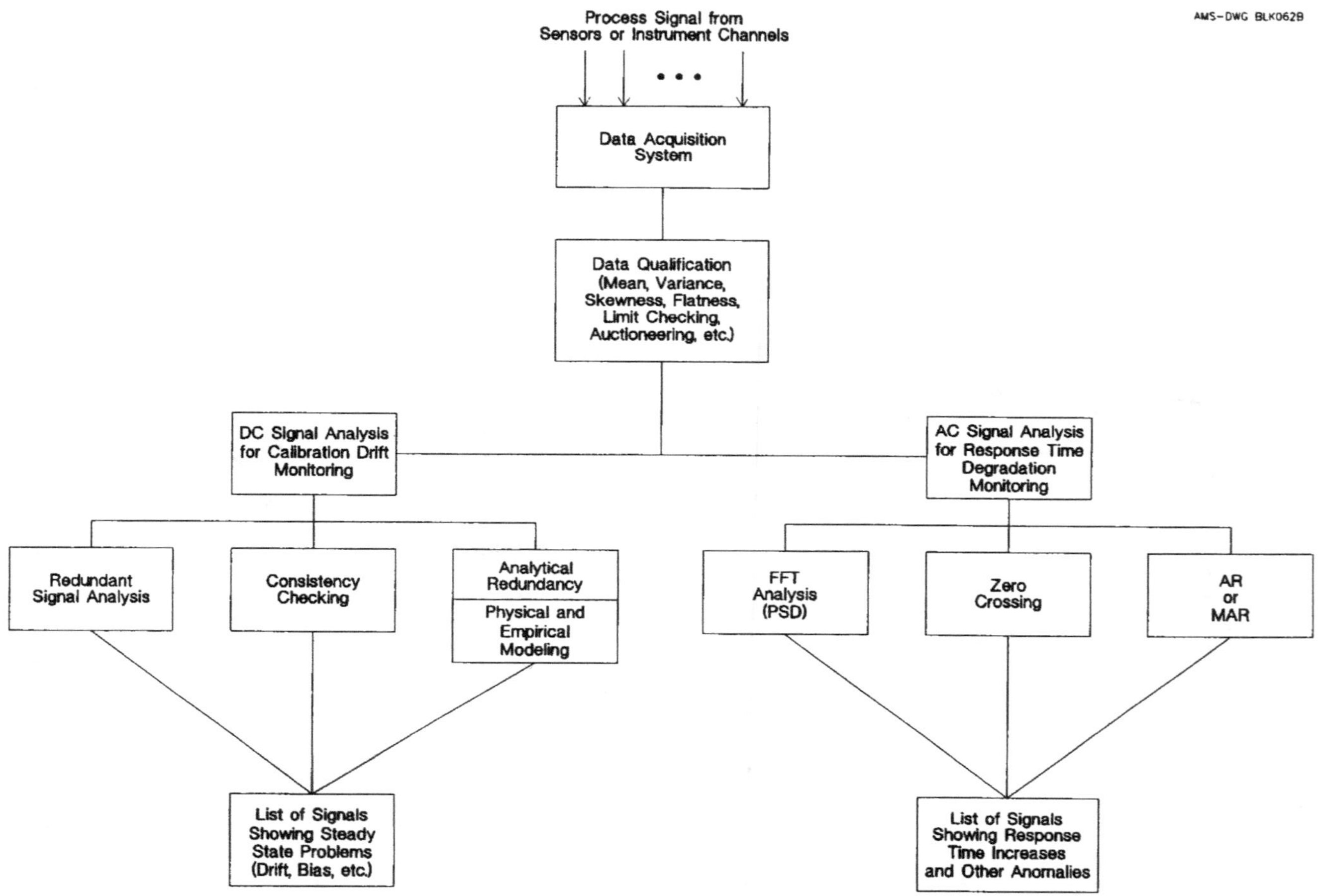

Figure 5.16 Data qualification and data analysis techniques for on-line verification of performance of installed instruments

is digitized and subtracting it from the noise record. Figure 5.17 illustrates the principle of the zero crossing analysis. It is apparent that if the instrument becomes sluggish, the number of zero crossings in a given period of time will decrease provided that the characteristics of the input noise remains the same. Therefore, this method can be used to detect relative changes in the response time of an instrument. Figure 5.18 shows zero crossing results for a number of flow sensors tested in our laboratory test loop. The results are shown from tests performed on two separate occasions to demonstrate the repeatability of the zero crossing tests. Figure 5.19 shows how the zero crossing rate for a Rosemount pressure transmitter decreases as the response time of the transmitter is increased by increasing the damping in the sensor electronics.

For certain dynamic systems, it has been shown[13] that under appropriate conditions, the zero crossing rate (Z) of an instrument that is driven by random noise is related to response time (τ) by the following equation:

$$\tau = \frac{C}{Z} \qquad (5.14)$$

where C is a proportionality constant unique to the instrument. The drawback of Equation 5.14 is that the constant C is not generally known and must be identified for each sensor or signal, and this is usually a formidable task. Nevertheless, the zero crossing approach is useful if a threshold is established for the zero crossing rate of an instrument and used to identify changes.

Univariate Autoregressive Modeling

The noise record can be fit to a univariate Autoregressive model (AR) of the following form to provide the step response of the instrument.

$$x(t) = \sum_{i=1}^{n} a_i \, (t - i\Delta T) + V(t) \qquad (5.15)$$

where $x(t)$ is the noise record, $V(t)$ is a forcing function, and a_i's are the parameters of the AR model. These parameters are determined by fitting the noise record to Equation 5.15. Once the AR parameters are known, the dynamic characteristics of the sensor can be identified and monitored for changes. Figure 5.20 shows the PSD of data for a pressure sensor with a clear and partially blocked sensing line. The sensing line was intentionally blocked in laboratory experiments to demonstrate that the AR model can clearly show the resulting change in the PSD of the sensor noise output.

FFT Analysis

The noise data may be analyzed in the frequency domain by a Fast Fourier Transformation (FFT) of the noise data to provide the PSD of the sensor. The PSD is then fit to a mathematical model (e.g., transfer function) that describes the dynamic behavior of the sensor, and the response time is obtained from the model. For qualitative tests, the PSD is plotted and its shape is monitored for changes. Figure 5.21 shows two PSDs for a Barton pressure transmitter. A normal and a degraded behavior is shown. The degradation was induced to demonstrate how it affects the PSD of the sensor.

Multivariate Autoregressive Model

As in the case of calibration drift monitoring, sensor response time degradation monitoring must incorporate a means to determine if a change in sensor dynamics is due to the sensor or the process. A sophisticated method for separating sensor effects from process effects called Multivariate Autoregressive (MAR) modeling is available, in which, redundant noise signals are sampled simultaneously and the common components of the signals that are due to the process are separated from the rest of the signal. The potential of this method will be explored in Phase II.

5.7   Pattern Recognition, Neural Networks, and Expert Systems

Pattern recognition and neural network techniques can be used for data qualification, calibration drift monitoring, and response time degradation testing. For data qualification and drift monitoring, normal APD and signal deviation patterns at a given reactor condition are learned and compared with future patterns to detect departures from normal behavior. In the same manner, for sensor response time degradation monitoring, the shape of normal PSDs when the sensors are new are learned and used to identify changes as the sensors age in the process. Figure 5.22 shows how the shape of noise PSDs corresponds to the dynamic response of a group of sensors. A data base can be created with various PSDs for new sensors and used to determine if future PSDs deviate from the PSDs in the database.

The pattern recognition and neural network techniques will be incorporated into an expert system to be developed in Phase II. The expert system will use artificial intelligence and reasoning techniques to identify and isolate sensor drift, response time degradation, and other faults.

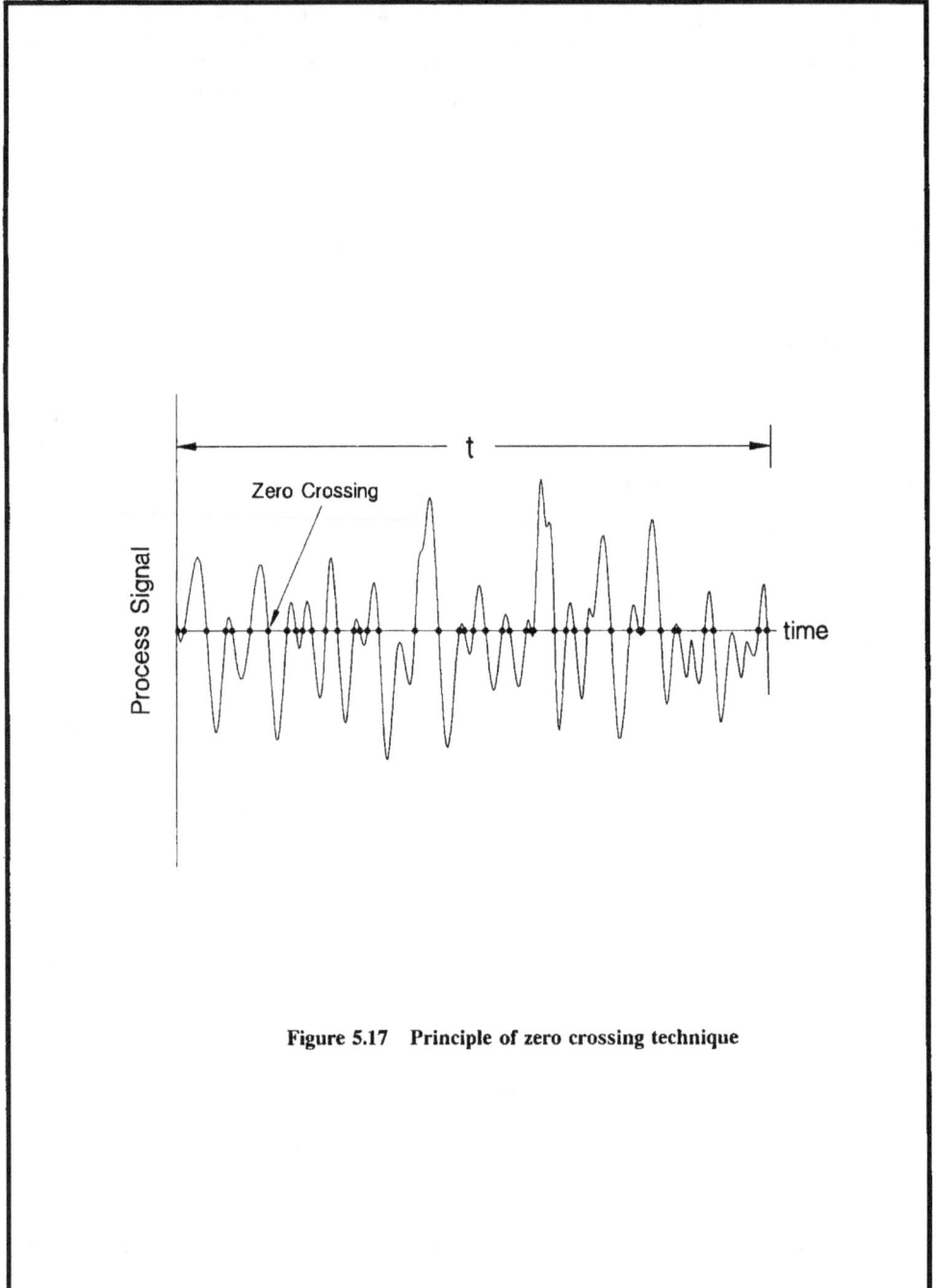

Figure 5.17   Principle of zero crossing technique

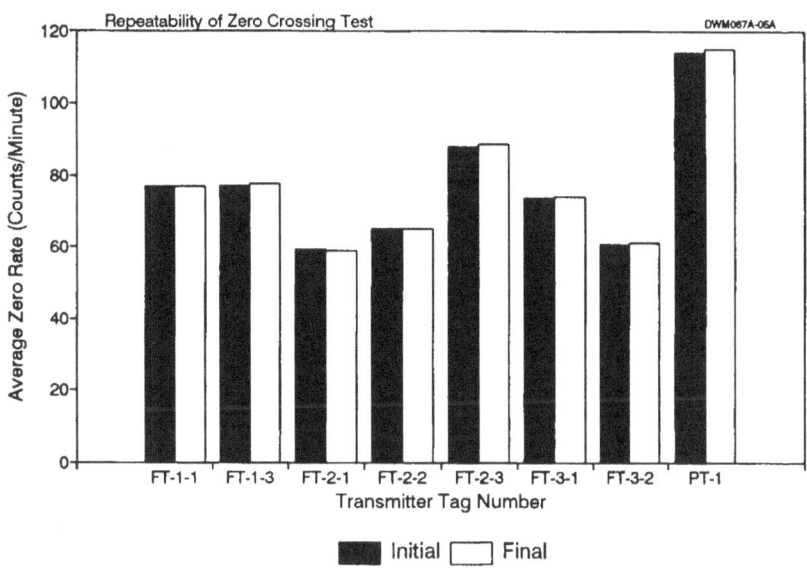

Figure 5.18 **Repeatability of zero crossing results for eight pressure transmitters tested in the laboratory test loop**

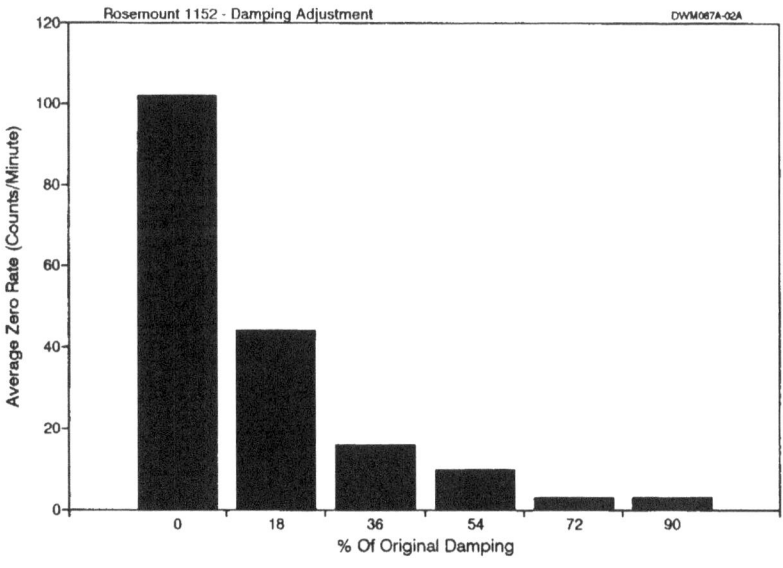

Figure 5.19 **Zero crossing rate as a function of damping for a Rosemount pressure transmitter**

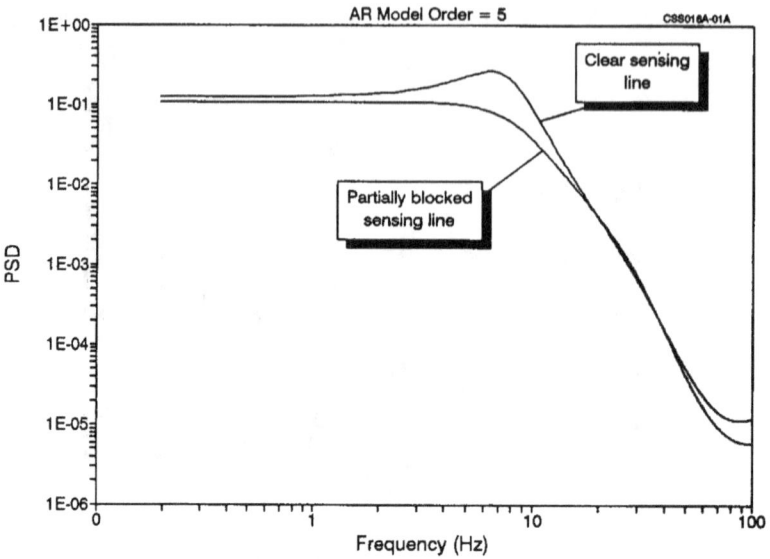

Figure 5.20   Effect of sensing line blockage on the PSD of a pressure
transmitter from AR model

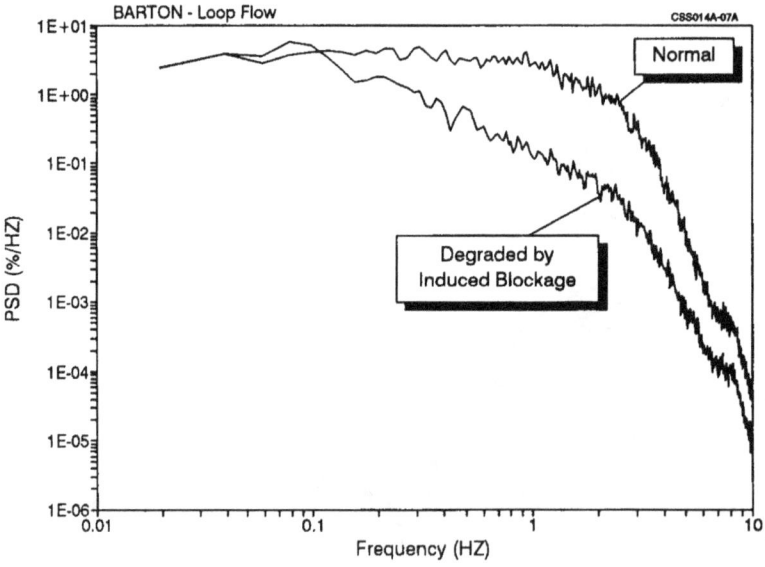

Figure 5.21   Detection of response time degradation by FFT analysis

AMS-DWG N01009E

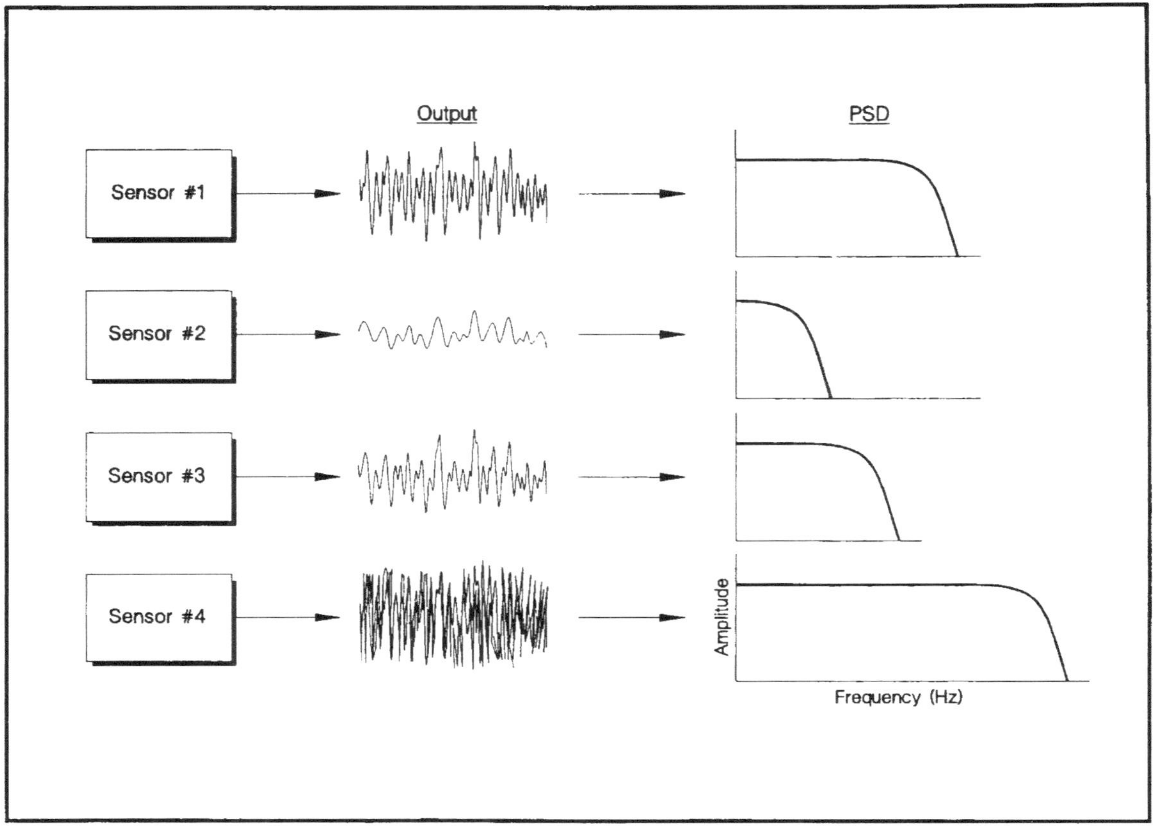

Figure 5.22 Illustration of raw noise data and corresponding PSD patterns

6. On-Line Testing of Calibration of Temperature Sensors in PWRs

Of all the methods that have been researched in the last ten years, a method called "cross calibration" has been the most successful method to date in providing on-line calibration testing capability to the nuclear power industry. The method is currently used in many pressurized water reactors (PWRs) to remotely test the calibration of the primary coolant RTDs as installed in the plant. Without additional effort, the core exit thermocouples can be included in the cross calibration process. However, since RTDs are generally more accurate than thermocouples, the RTDs must be used exclusively or with a large weighting factor to give an estimate of the reactor coolant temperature. The test is performed from the control room area where the field leads from the temperature sensors reach their signal conversion and signal conditioning equipment in the plant protection system cabinets.

The success of the cross calibration method in PWRs is due mainly to a large number of redundant temperature sensors that are measuring the same temperature when the plant is at isothermal conditions. As in most on-line calibration testing techniques, the cross calibration method is susceptible to common mode problems meaning that it may not detect a common bias if it exists in a group of RTDs or thermocouples that are cross calibrated together. There are two remedies to this problem: 1) include a freshly calibrated RTD in each set of cross calibration, 2) use the analytical redundancy technique to obtain an independent estimate for the reactor coolant temperature to ensure that the RTDs or thermocouples have not drifted together in the upward or downward direction. The first remedy is more effective and is probably the only method that will give unquestionable results, but it is more inconvenient and expensive to practice than the second remedy. The problem with the second remedy is that its uncertainties may be larger than the bias that we may be trying to resolve.

There are simpler ways than analytical redundancy to obtain an independent estimate of the reactor coolant temperature, but the simpler methods usually have larger uncertainties. For example, at isothermal conditions, the steam pressure at the steam generator outlet can be measured and used with steam tables to provide an estimate of the steam generator temperature and the primary coolant water that circulates in the steam generator. The problem with this method is that there are uncertainties in every step of the way starting

with the uncertainties associated with measurement of the steam pressure.   The combination of the uncertainties often makes it difficult to obtain a reliable estimate of the temperature to better than 0.5°F which is usually needed for RTD cross calibration tests in PWRs.

The cross calibration method is described in detail in the following section.

6.1 Cross Calibration Principle

The reactor coolant temperature in a PWR is measured with twenty to forty RTD elements depending on the plant. Figure 6.1 shows two loops of a four-loop PWR with the RTDs that are installed in the hot leg and cold leg pipes. At isothermal conditions, all RTDs should be at the same temperature and can therefore be compared to identify the outliers. To perform a cross calibration test, the resistances of all RTDs are measured at isothermal conditions and converted to corresponding temperatures using the most recent calibration tables for the RTDs. The temperatures are then averaged and the deviation of each individual RTD from the average temperature is calculated (Figure 6.2). Any RTD that exceeds a pre-specified deviation is flagged and/or removed from the average and the process is repeated as necessary to identify the outliers (if any). Figure 6.3 shows a flow chart of steps that are taken for the acquisition of a set of cross calibration data.

The cross calibration test can be performed at one or more temperature plateaus at isothermal conditions when the reactor coolant hot leg and cold leg temperatures are at approximately the same temperature. The test can also be performed during plant startup or shutdown when the temperature is increasing or decreasing monotonically at a reasonably constant rate.

Until recently, cross calibration was thought to be limited to the test of consistency between a group of RTDs. However, recent research[8] has proven that the drift of current generation of nuclear grade RTDs is random rather than systematic, and therefore, the accuracy of cross calibration results can be made to approach that of a laboratory calibration, especially if one or more newly calibrated RTDs are included in the cross calibration process.

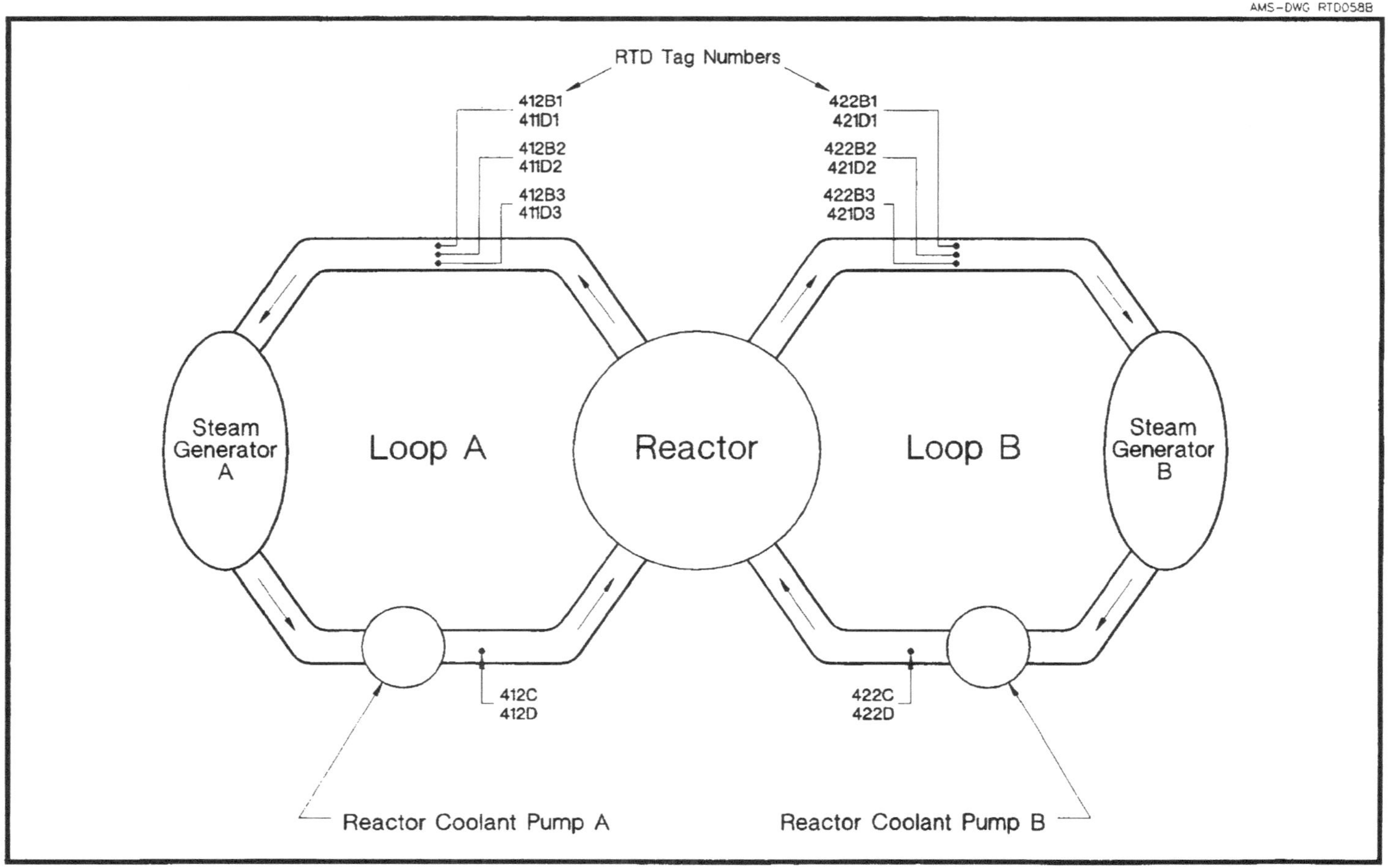

Figure 6.1   Simplified diagram of two loops of a four-loop PWR showing the hot leg and cold leg RTDs

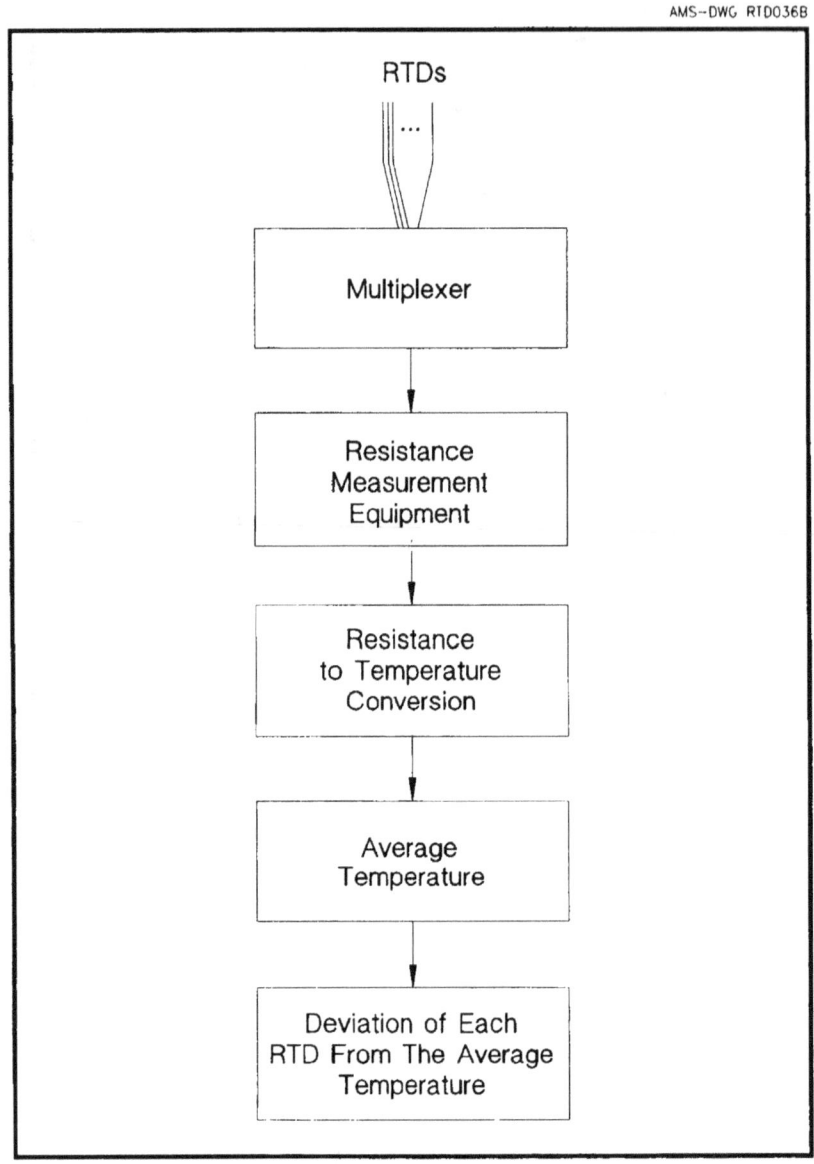

Figure 6.2   Illustration of cross calibration procedure

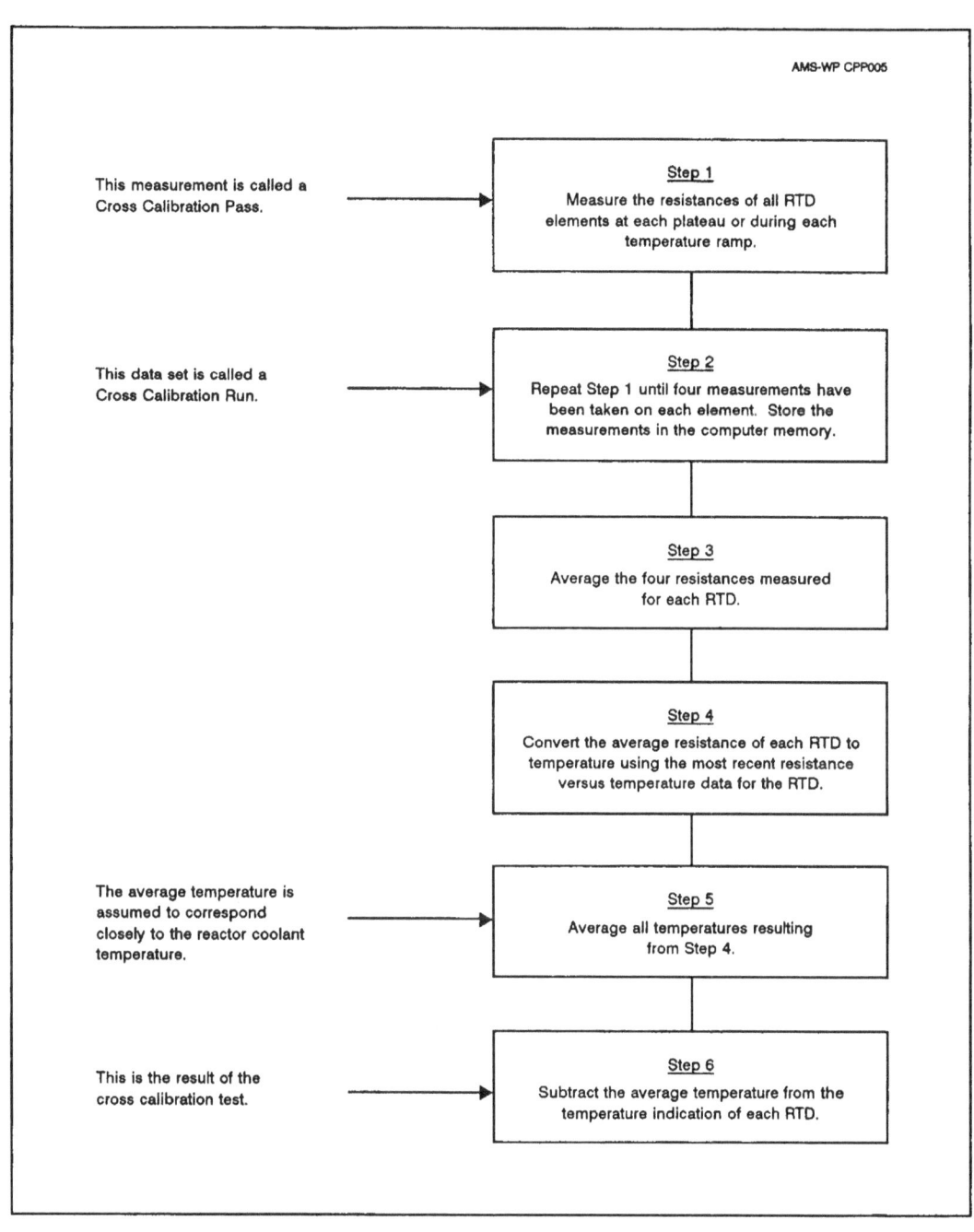

This measurement is called a
Cross Calibration Pass.

Step 1

Measure the resistances of all RTD
elements at each plateau or during each
temperature ramp.

This data set is called a
Cross Calibration Run.

Step 2

Repeat Step 1 until four measurements have
been taken on each element. Store the
measurements in the computer memory.

Step 3

Average the four resistances measured
for each RTD.

Step 4

Convert the average resistance of each RTD to
temperature using the most recent resistance
versus temperature data for the RTD.

The average temperature is
assumed to correspond
closely to the reactor coolant
temperature.

Step 5

Average all temperatures resulting
from Step 4.

This is the result of the
cross calibration test.

Step 6

Subtract the average temperature from the
temperature indication of each RTD.

Figure 6.3   Data acquisition and data analysis steps in RTD cross calibration

6.2   Cross Calibration Test Equipment and Procedure

The cross calibration tests are performed using a sensor scanning system involving a computer, a multiplexer, and a precision digital multimeter (Figure 6.4). The data for RTDs are collected and analyzed according to the following procedure:

1. Sequence through all RTDs and measure the element resistances.

2. Repeat Step 1 to obtain four passes.

3. Average the four resistance measurements for each RTD element.

4. Convert the average resistance of each element to its equivalent temperature using the latest resistance-versus-temperature relationship for the RTD.

5. Average the temperatures identified in Step 4. This is the best estimate of the plant's primary coolant temperature under the conditions tested.

6. Subtract the average temperature identified in Step 5 from the temperature indications of each RTD element. This is referred to as the deviation of each RTD and is denoted as ΔT.

7. Scan the ΔT column and flag those RTDs that exceed a predetermined criteria (e.g., ± 0.3°F).

8. If the deviation of any RTD element exceeds a predetermined rejection criteria (e.g., ± 1.0°F), remove the element's measurement from the data and repeat from Step 5 to obtain a new average temperature.

9. Repeat Step 8 until all RTD elements which have ΔTs greater than the rejection criteria have been eliminated from the average.

The result of this procedure is referred to as a cross calibration run. A typical cross calibration run for 24 RTD elements in a PWR is shown in Table 6.1. The cross calibration tests are usually performed at 3 to 5 temperature plateaus and 5 to 15 runs are often collected at each plateau depending on the plant's temperature stability. The results of the repeated runs are then averaged to obtain a best estimate of the deviation of each RTD. Table 6.2 shows the results of

a cross calibration test involving 14 runs on 24 RTD elements.

The cross calibration test requires access to all RTDs at essentially the same time. This is accomplished by a variety of methods depending on the plant. These methods are:

1. Disconnect the RTD field leads from the signal conversion equipment and connect them to the cross calibration test equipment. With this method, the plant operators would have to depend on other temperature sensors for indication. In most plants, there are provisions (switches or relays) to disconnect the RTDs without a need to physically lift any leads. This is explained below.

2. In plants where the RTDs can be disconnected by electronic relays, a trigger signal is sent from the test equipment to the relay to disconnect the RTD from the rest of the instrument channel and read its resistance (Figure 6.5). The trigger signal is then removed, the RTD is restored to its original configuration, and the process is continued with the next RTD. With this method, only one RTD is out of service at a time and while the RTD is being tested, a fixed resistor is automatically substituted for the RTD to keep the RTD channel from displaying a trip signal or alarm.

3. With a digital instrumentation system, such as the Westinghouse Eagle 21, the resistance of the RTDs can be measured passively while the RTDs are in normal service. The voltage drop across and the current flow through each RTD element is measured and divided to obtain the RTD resistance (Figure 6.6).

6.3   Validation of Cross Calibration Approach

As a part of the development of the cross calibration method under an earlier R&D project for NRC[8], a systematic program of laboratory tests were carried out to demonstrate the validity and determine the accuracy of the cross calibration technique. The program involved a number of nuclear-grade RTDs which were tested in a constant temperature bath at 572°F (300°C). The results are shown in Table 6.3 along with a measurement of the true temperature of the bath measured with a Standard Platinum Resistance Thermometer (SPRT). It is apparent that the average

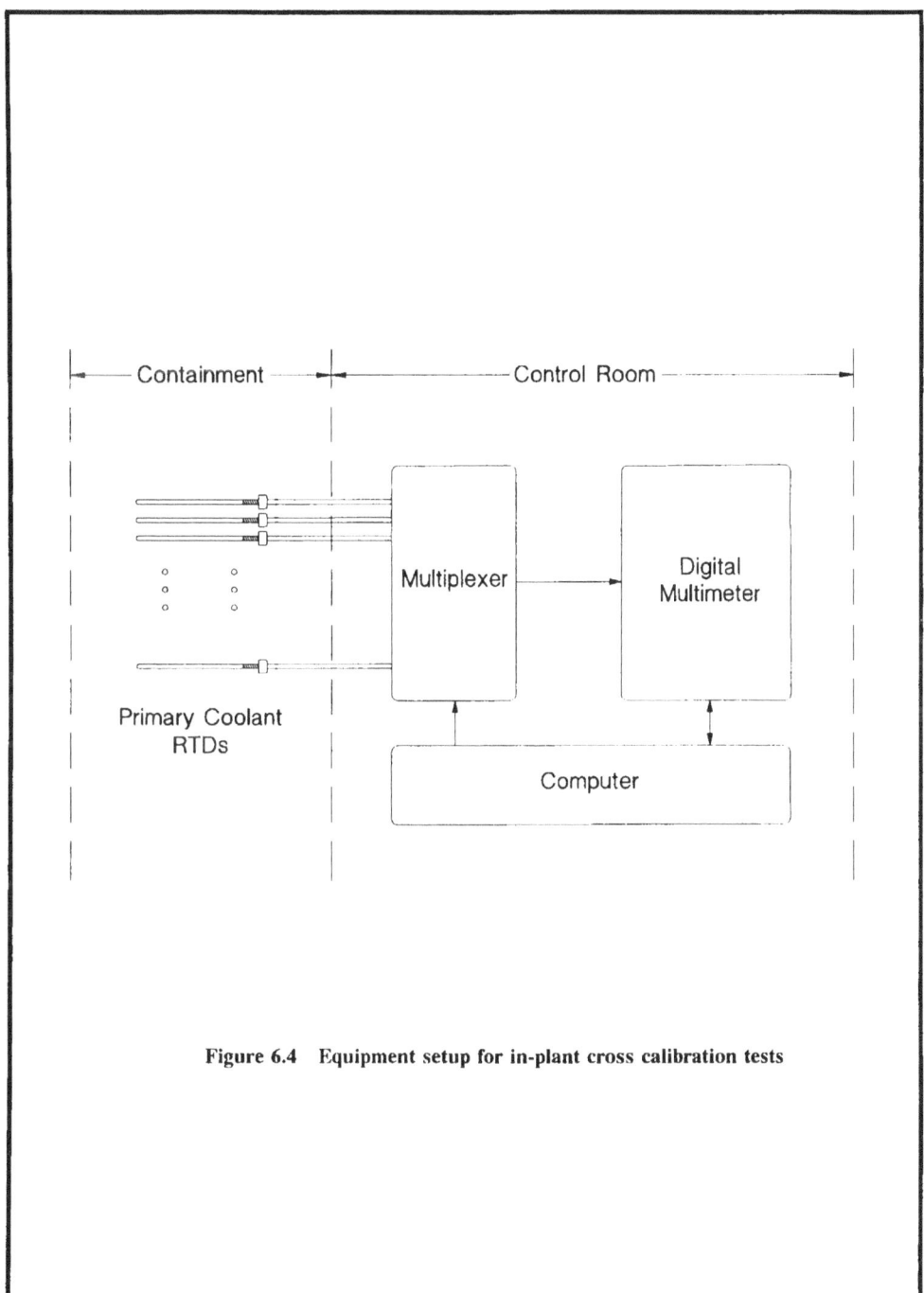

Figure 6.4   Equipment setup for in-plant cross calibration tests

Table 6.1

RTD Cross Calibration
Version #: 4.1
Method: Plateau

Performed: 10-24-1991   19:26:46   Temperature Plateau: 547°F

Tag No.	Resistance Measurements (Ohms)				Average Resistance Ohms	Temp. (°F)	Deviation ΔT (°F)
	Pass 1	Pass 2	Pass 3	Pass 4			
412B1	414.3931	414.3903	414.3927	414.4591	414.4088	547.584	-0.033
411D1	414.3486	414.3498	414.3536	414.3666	414.3547	547.652	0.035
412B2	414.2208	414.2603	414.1944	414.1789	414.2136	547.750	0.133
411D2	414.3576	414.3614	414.3628	414.3784	414.3651	547.405	-0.212
412B3	414.1348	414.2221	414.2005	414.2076	414.1912	547.482	-0.134
411D3	414.3326	414.3348	414.3390	414.3554	414.3405	547.697	0.080
412C	414.4881	414.4846	414.4281	414.4469	414.4619	547.712	0.095
412D	414.2606	414.3330	414.3170	414.3040	414.3036	547.503	-0.114
422B1	414.2206	414.3127	414.2628	414.2336	414.2574	547.515	-0.102
421D1	414.2890	414.2892	414.2912	414.3088	414.2946	547.405	-0.211
422B2	414.3074	414.2777	414.2789	414.2962	414.2901	547.497	-0.120
421D2	414.0812	414.0804	414.0826	414.0994	414.0859	547.455	-0.162
422B3	414.0388	414.0337	413.9802	414.0001	414.0132	547.343	-0.274
421D3	414.1890	414.1886	414.1930	414.2082	414.1947	547.544	-0.073
422C	414.4119	414.4104	414.4165	414.4328	414.4179	547.921	0.304#
422D	414.3634	414.3331	414.3141	414.3743	414.3462	547.860	0.243
432B1	414.0776	414.0776	414.0789	414.1083	414.0856	547.625	0.008
431D1	414.1074	414.1100	414.1162	414.1300	414.1159	547.445	-0.172
432B2	414.5670	414.5613	414.5713	414.5247	414.5561	547.989	0.372#
431D2	414.1730	414.1748	414.1814	414.1950	414.1811	547.574	-0.042
432B3	414.1224	414.1549	414.1458	414.1571	414.1450	547.699	0.082
431D3	414.1728	414.1752	414.1816	414.1966	414.1816	547.603	-0.014
432C	414.4862	414.4660	414.4519	414.5231	414.4818	547.884	0.267
432D	414.5917	414.5764	414.5867	414.6010	414.5889	547.662	0.045

Average Temperature: 547.617

Notes: * Not Used in Average   # Deviation Limit Exceeded

Table 6.2

Deviation of RTDs in Repeated Runs at the 547°F Plateau

Tag No.	Deviation (°F) Per Run Number														Average Deviation (°F)
	1	2	3	4	5	6	7	8	9	10	11	12	13	14	
412B1	-0.03	-0.04	-0.02	0.04	-0.02	0.07	0.05	0.02	0.08	-0.05	0.00	-0.01	0.01	0.01	0.01
411D1	0.04	0.05	0.06	0.09	0.08	0.07	0.03	0.06	0.05	0.07	0.08	0.11	0.09	0.08	0.07
412B2	0.14	0.02	0.04	0.10	0.14	0.03	0.09	0.05	0.09	0.14	0.08	0.07	0.10	0.09	0.08
411D2	-0.21	-0.20	-0.18	-0.14	-0.15	-0.18	-0.21	-0.18	-0.19	-0.18	-0.17	-0.13	-0.14	-0.16	-0.17
412B3	-0.13	-0.24	-0.21	-0.15	-0.19	-0.12	-0.23	-0.24	-0.19	-0.24	-0.15	-0.17	-0.18	-0.19	-0.19
411D3	0.08	0.10	0.11	0.15	0.14	0.12	0.07	0.10	0.11	0.11	0.14	0.16	0.14	0.13	0.12
412C	0.10	0.09	0.08	0.18	0.14	0.06	0.10	0.16	0.18	0.15	0.03	0.16	0.25	0.20	0.14
412D	-0.11	-0.15	-0.03	-0.05	0.00	-0.17	-0.09	-0.07	-0.15	-0.18	-0.20	-0.10	0.01	-0.05	-0.10
422B1	-0.10	-0.09	-0.14	-0.18	-0.13	-0.21	-0.17	-0.18	-0.13	-0.08	-0.20	-0.17	-0.09	-0.12	-0.14
421D1	-0.21	-0.20	-0.19	-0.18	-0.19	-0.21	-0.21	-0.18	-0.20	-0.18	-0.20	-0.16	-0.14	-0.17	-0.19
422B2	-0.12	0.01	-0.02	-0.05	-0.09	-0.09	-0.06	-0.08	-0.07	-0.01	-0.07	-0.06	-0.02	-0.06	-0.06
421D2	-0.16	-0.15	-0.14	-0.11	-0.14	-0.17	-0.16	-0.13	-0.14	-0.13	-0.15	-0.10	-0.07	-0.11	-0.13
422B3	-0.27	-0.27	-0.24	-0.29	-0.33	-0.33	-0.36	-0.31	-0.33	-0.27	-0.24	-0.32	-0.23	-0.27	-0.29
421D3	-0.07	-0.07	-0.06	-0.03	-0.05	-0.08	-0.08	-0.04	-0.05	-0.05	-0.06	-0.02	0.02	-0.03	-0.05
422C	0.30	0.39	0.24	0.19	0.20	0.31	0.35	0.34	0.22	0.33	0.25	0.20	0.16	0.26	0.27
422D	0.24	0.22	0.20	0.10	0.07	0.21	0.28	0.30	0.14	0.20	0.14	0.17	0.03	0.15	0.18
432B1	0.01	0.05	0.05	0.06	0.01	0.12	0.04	0.01	0.06	-0.01	0.05	0.00	-0.03	0.00	0.03
431D1	-0.17	-0.17	-0.20	-0.21	-0.20	-0.16	-0.19	-0.19	-0.19	-0.17	-0.15	-0.18	-0.22	-0.20	-0.18
432B2	0.37	0.26	0.22	0.25	0.38	0.32	0.33	0.35	0.34	0.30	0.33	0.25	0.24	0.27	0.30
431D2	-0.05	-0.05	-0.07	-0.09	-0.08	-0.03	-0.06	-0.06	-0.06	-0.05	-0.03	-0.05	-0.10	-0.08	-0.06
432B3	0.08	0.07	0.12	0.03	0.03	0.02	0.14	0.06	0.04	0.06	0.08	0.10	0.00	0.03	0.06
431D3	-0.02	-0.02	-0.06	-0.08	-0.08	-0.02	-0.04	-0.03	-0.05	-0.03	-0.02	-0.04	-0.09	-0.06	-0.04
432C	0.26	0.20	0.31	0.27	0.30	0.29	0.26	0.19	0.34	0.20	0.34	0.23	0.21	0.22	0.26
432D	0.04	0.18	0.15	0.10	0.15	0.15	0.11	0.06	0.08	0.05	0.11	0.06	0.06	0.08	0.10

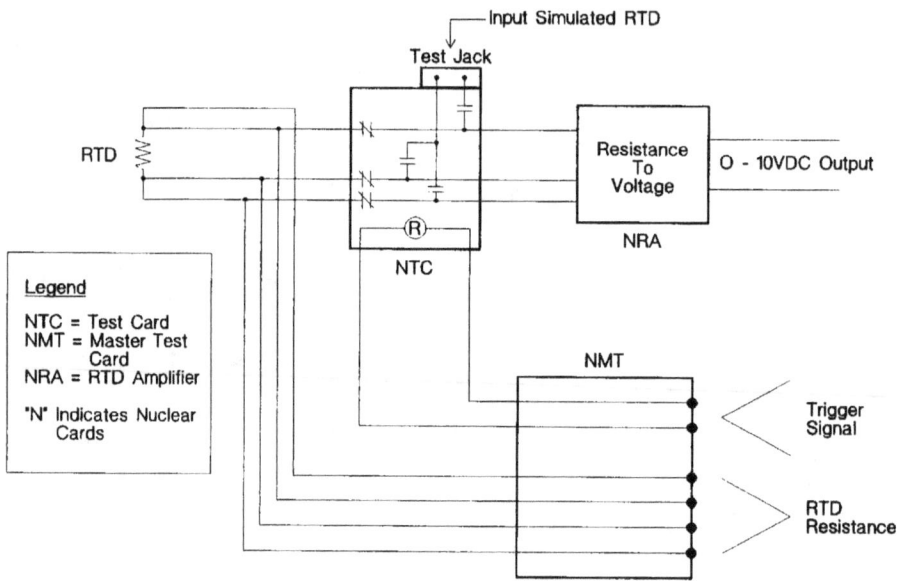

Figure 6.5   Westinghouse 7300 System test cards involved in RTD
              cross calibration

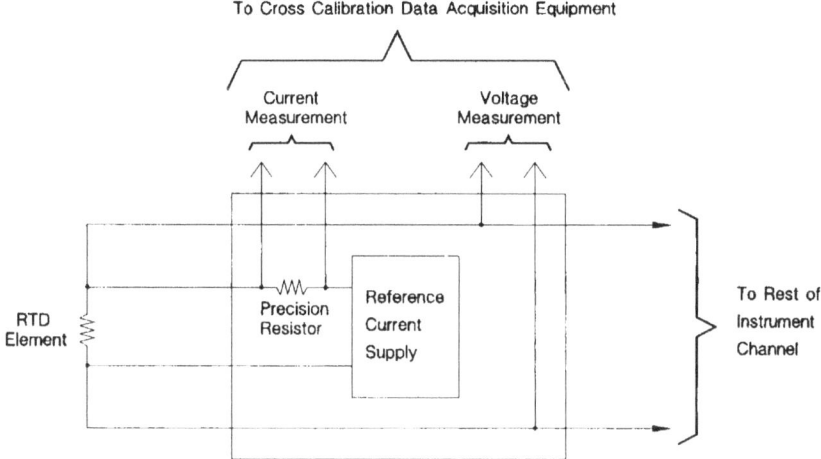

Figure 6.6   Cross calibration test of RTDs in Eagle 21 System

Table 6.3

Laboratory Validation of
Cross Calibration of RTDs

300°C Plateau

Tag	Resistance Measurements (Ohms)				Avg. Res. (Ohms)	Temp. (°C)	Dev. (°C)
	Pass 1	Pass 2	Pass 3	Pass 4			
21	212.2644	212.2643	212.2741	212.2634	212.2666	*303.537	# 2.751
19	429.4928	429.4910	429.5060	429.4940	429.4960	300.759	-0.028
12A	429.1156	429.1152	429.1298	429.1134	429.1185	300.811	0.025
12C	429.0084	429.0088	429.0224	429.0102	429.0125	300.807	0.021
03	424.6416	424.6324	424.6384	424.6494	424.6405	300.785	-0.001
18	429.8286	429.8200	429.8228	429.8394	429.8277	300.819	0.033
15A	424.6800	424.6818	424.6862	424.6988	424.6867	300.774	-0.012
15C	424.2652	424.2698	424.2740	424.2818	424.2727	300.774	-0.012
13A	428.9974	428.9962	429.0018	429.0144	429.0025	300.762	-0.024
13C	428.9734	428.9702	428.9726	428.9840	428.9751	300.771	-0.015
9C	430.2376	430.2382	430.2478	430.2404	430.2410	300.787	0.001
9A	430.1488	430.1532	430.1658	430.1480	430.1540	300.780	-0.006
17A	424.3216	424.3148	424.3264	424.3152	424.3195	300.840	0.054
17C	424.2266	424.2230	424.2322	424.2186	424.2251	300.833	0.047
16A	424.7866	424.7882	424.7912	424.7800	424.7865	300.806	0.020
16C	424.5208	424.5294	424.5258	424.5228	424.5247	300.806	0.020
07	430.0636	430.0732	430.0620	430.0628	430.0654	300.724	-0.062
20	430.3424	430.3492	430.3344	430.3448	430.3427	300.749	-0.037
SPRT-1	54.7764	54.7778	54.7761	54.7767	54.7768	300.788	0.002

Average Temperature = 300.786°C

* = Not Used in Average   # = Deviation Limit Exceeded

- 62 -

temperature, indicated by this group of RTDs is a pretty accurate estimate of the true temperature of the bath as evident by the close agreement between the reading of the SPRT and the average of the RTDs. The same experiment was conducted on a group of thermocouples (Table 6.4). Although the indication of the individual thermocouples have large deviations from the average of the group, the reading of the SPRT is close to the average of the thermocouples. That is, the average indication of a large group of reasonable sensors can be assumed to closely represent the true temperature of the process in which the sensors are used.

6.4 Accuracy of Cross Calibration Results

The accuracy of cross calibration results depends on the accuracy of the RTD calibration charts, precision and accuracy of the resistance measurement equipment, the stability and uniformity of the plant temperatures during the tests, the number of sensors that are cross calibrated together, the number of passes in each cross calibration run, and the number of runs in a cross calibration test. The most important of these factors and the most difficult to quantify are the plant temperature stability and uniformity errors. These errors and the methods to identify them are described in NUREG/CR-5560.[8]

6.5 Cross Calibration Tests at Power

The best time to perform a cross calibration test is when the plant is at isothermal conditions giving 20 to 40 redundant RTDs to be intercompared. As mentioned earlier, the test can also be performed under temperature ramp conditions during startup or shutdown if the ramp rate is sufficiently constant and has a suitable rate.

In addition to isothermal and temperature ramp conditions, the cross calibration tests may be performed at normal power operation, but the accuracy of the results will not be as good as when the tests are performed at isothermal conditions. At power, there are usually three or four redundant RTD elements in each hot leg, and one or two elements in each cold leg. The hot leg RTDs in each loop can be cross calibrated against one another, using the other hot leg and cold leg RTDs in the plant to help verify that: 1) the hot leg RTDs under test are exposed to stable and uniform temperatures, and 2) an independent estimate of the temperature of the hot leg loop being tested is obtained.

Cross calibration tests have been successfully performed in PWRs under normal operating conditions and the results independently verified. However, this does not mean that the tests will always be successful at power. With as little as four redundant RTDs, each case must be treated differently and all means must be used to minimize the uncertainties of the tests.

Figure 6.7 shows a plot of the deviations of three hot leg RTDs as a function of power in a four-loop PWR. The deviations were obtained from cross calibration data obtained continuously from 40 percent to 100 percent power. It is apparent that the deviations increase with power due to increased temperature stratification in the hot leg pipe as the power is increased. As shown in the figure, the deviations are extrapolated to zero power and the results were compared with the tests at isothermal conditions. This comparison showed reasonable agreement verifying that this cross calibration test was successfully implemented at power.

It should be pointed out that since the RTD deviations in Figure 6.7 are calculated with respect to their own average, the magnitude of the deviations may not correspond to the actual magnitude of the temperature stratification.

Table 6.4

Laboratory Validation of Cross Calibration of Thermocouples

I.D.	Type	Thermocouple Output (millivolt)					Temp. (°C)	Dev. (°C)
		Pass 1	Pass 2	Pass 3	Pass 4	Avg		
1	K	8.103	8.103	8.103	8.103	8.103	199.16	-1.17
2	K	8.116	8.115	8.115	8.115	8.115	199.46	-0.87
3	K	8.201	8.201	8.201	8.201	8.201	201.61	1.28
4	K	8.123	8.124	8.130	8.125	8.126	199.74	-0.59
5	E	13.471	13.471	13.471	13.470	13.471	200.71	0.38
6	E	13.500	13.500	13.499	13.499	13.500	201.10	0.77
7	E	13.513	13.514	13.510	13.512	13.512	201.26	0.93
8	E	13.433	13.442	13.430	13.420	13.431	200.17	-0.16
9	J	10.755	10.758	10.757	10.757	10.758	199.67	-0.66
10	J	10.831	10.834	10.834	10.833	10.834	201.04	0.71
11	J	10.725	10.725	10.725	10.725	10.725	199.07	-1.26
12	J	10.830	10.829	10.830	10.829	10.830	200.96	0.63
SPRT	Ω	45.334	45.334	45.334	45.334	45.334	200.38	0.05

Average Temperature = 200.33°C

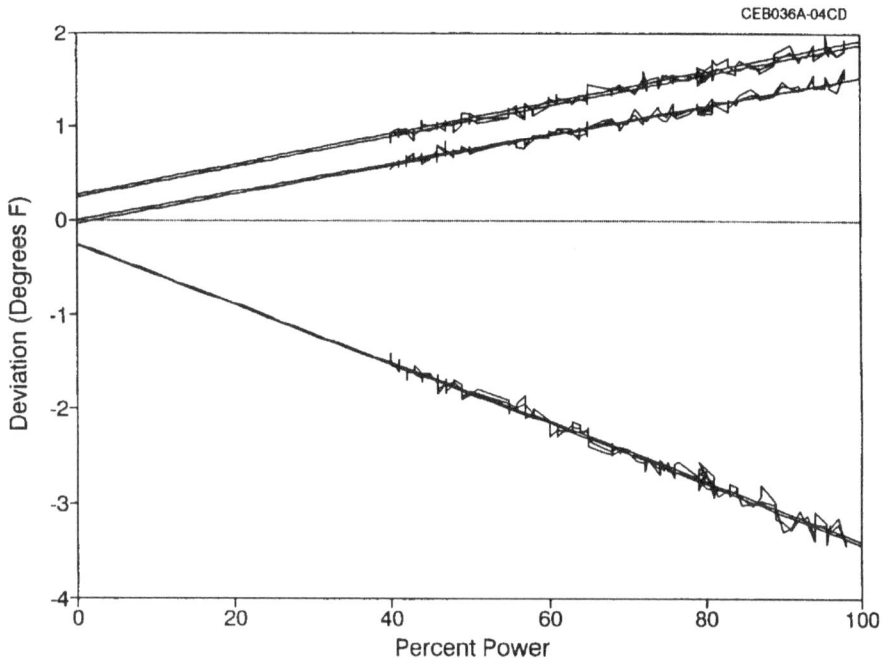

Figure 6.7   Results of cross calibration test of three hot leg RTDs at power

7. On-Line Testing of Calibration of Pressure Transmitters

The cross calibration method for on-line testing of calibration of temperature sensors in nuclear power plants was described in detail in Chapter 6. The cross calibration method works very well for PWR temperature sensors when the plant is at isothermal conditions, providing 20 to 40 redundant temperature elements to be intercompared. This level of redundancy is not normally available for other sensors. For example, the maximum number of redundant pressure transmitters in a nuclear power plant is typically four.

To make up for the lack of redundancy of pressure transmitters, instead of obtaining a few snapshots of the sensor readings, as is done in RTD cross calibration, the transmitter outputs should be continuously recorded over an entire fuel cycle and examined periodically to determine drift and other problems. In addition, several independent methods should be used simultaneously to account for common mode drift and provide added confidence in the results. Figure 7.1 shows a block diagram of how an on-line monitoring system can use all the available tools to help in on-line testing of calibration of pressure transmitters.

Presently, there is no automated means of examining the on-line monitoring data, nor is it time yet to develop such an automated system. The method is not fully validated, and there is no way to prove to date that the accuracy of on-line calibration testing methods will be sufficient enough to detect a drift of as little as a few tenths of a percent in the output of a transmitter. Table 7.1 shows typical acceptance criteria for calibration of representative pressure transmitters in nuclear power plants.

The on-line monitoring results are presently examined visually using a set of rules. The rules for a group of four redundant transmitters for example are:

1. If none of the four transmitters have drifted, then only one transmitter is calibrated to rule out common mode drift. This approach reduces the calibration effort by 75 percent.

2. If one of the four redundant transmitters has drifted but the remaining three are stable (Figure 7.2), then the transmitter that has exhibited the drift and one of the stable transmitters are calibrated. This provides a 50 percent reduction in the calibration effort because only two transmitters are calibrated out of a redundant group of four.

The same type of rules or logic can be used in calibration reduction for a group of three transmitters. Figure 7.3 shows examples of drift scenarios for three redundant transmitters.

A few important questions must be answered before it can be determined whether an on-line monitoring system can provide adequate relief to the nuclear power industry. These questions are:

1. Is the method precise enough to detect a small drift to help avoid conventional calibrations?

2. What are the uncertainties of the method and if the uncertainties are combined, will it still leave enough margin for the method to be successful as a calibration reduction method?

3. Can common mode problems be effectively detected through analytical redundancy and other means?

4. If the method is not successful enough to replace conventional calibrations, will it still be useful for extending the instrument calibration and surveillance intervals? Can the method be used as a substitute for the monthly or quarterly surveillance tests?

5. If a sensor has not shown a significant drift, can we conclude that it is still in calibration?

6. What are the advantages of on-line monitoring techniques over the conventional calibration and response time testing techniques? A few examples of the advantages of on-line testing techniques are:

   a) Provides early warning of drift and incipient failures such as the oil loss syndrome in Rosemount pressure transmitters.[14]

   b) Provides the actual in-service performance of the instruments by accounting for the effects of installation and process operating conditions.

   c) Provides a means to include the sensors in the surveillance tests performed in between outages.

   d) Provides additional diagnostic capabilities beyond the capabilities of conventional calibrations.

   e) Reduces radiation exposure to the test personnel.

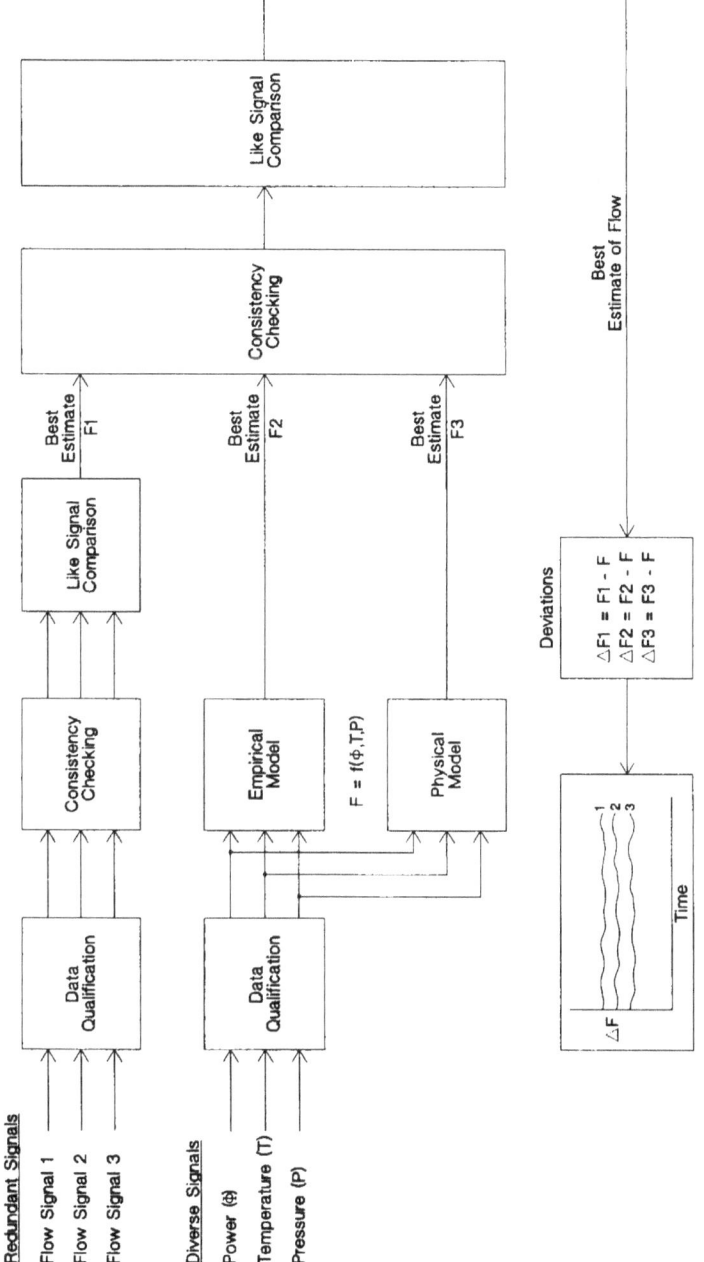

Figure 7.1   Illustration of an on-line drift monitoring system combining redundant and diverse signal analysis

Table 7.1

**Examples of Typical Acceptance Criteria for Calibration
of Nuclear Plant Pressure Transmitters**

Transmitter Manufacturer	Typical Acceptance Criteria (% of Span)
Barton	±0.5
Fischer & Porter	±0.5
Foxboro	±0.5
Rosemount	±0.25
Statham	±0.25
Tobar (Veritrak)	±0.5

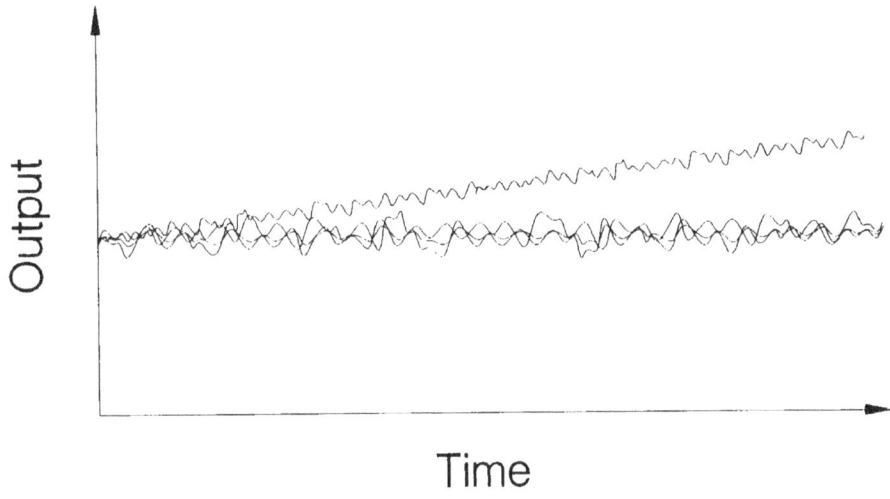

Figure 7.2 Illustration of a potential drift scenario for four redundant signals

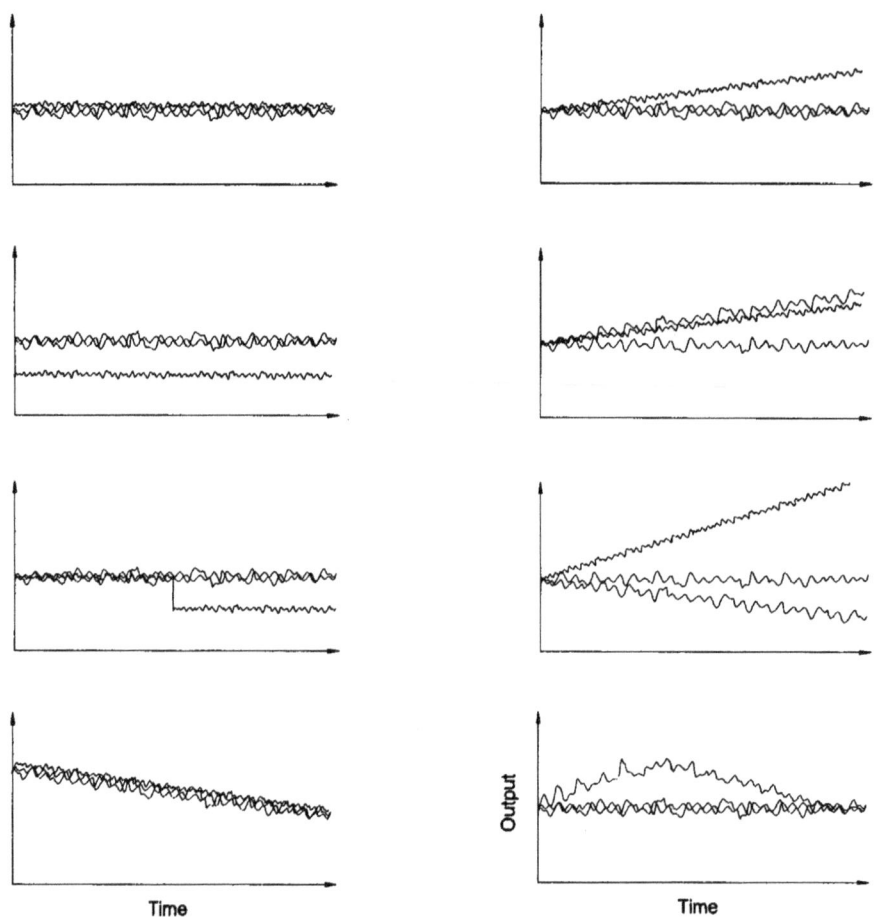

Figure 7.3 Examples of typical drift scenarios for three redundant signals

Can the above advantages outweigh the disadvantages of the on-line testing techniques and contribute to their success in receiving regulatory and industry approval?

The question of whether or not the on-line monitoring methods will be successful may have to be addressed on a case-by-case basis. A useful approach would be to produce a matrix of instruments and their potential faults versus the capabilities of the on-line monitoring techniques. The matrix can then be compared against the plant and NRC requirements to assess acceptability.

8. Phase I Accomplishments and Phase II Goals

The project reported herein is performed in two phases as follows:

- A Phase I feasibility study conducted over a nine month period.

- A Phase II research and development (R&D) project that began in October 1992 and is due for completion in the Fall of 1994.

In addition, a Phase III commercialization effort is currently underway to provide the results of the Phase I and Phase II developments to the nuclear power industry in terms of equipment, services, and training in the area of on-line testing of calibration and response time of process instrumentation channels. The Phase I and Phase II projects are partially funded by the NRC, but the Phase III project is an independent effort funded by AMS. This project is being conducted under a special U.S. government program that promotes the commercialization of federally-funded R&D projects.

The Phase I project has been successfully completed and the details are given in this report. In previous chapters, we covered the historical and technical background of the project. From here on out, we will devote the report to specific tasks completed in Phase I.

8.1 Phase I Accomplishments

The following tasks were successfully completed in Phase I.

Acquisition of Test Sensors and Associated Electronics

A number of RTDs, thermocouples, and pressure transmitters were obtained for the project from various sources. A partial listing of these sensors is given in Table 8.1 in terms of a tag number and the location of each sensor in our laboratory test loop. A unique tag number was assigned to each sensor to help track the performance of the sensor throughout the project. The list in Table 8.1 includes a number of smart sensors provided by Rosemount for evaluation in the Phase I project. These sensors were later purchased from Rosemount for continued evaluation in Phase II. In addition, arrangements have been made to obtain a number of nuclear and commercial grade RTDs and pressure transmitters from the Sandia National Laboratory for the Phase II project. These sensors have been used at Sandia for previous NRC projects and are no longer needed there.

Most of the sensors used in Phase I were available at AMS from previous R&D projects performed for the NRC and others. The remaining sensors were purchased through surplus sales from utilities or directly from regular suppliers. In addition, a Westinghouse Model 7300 instrumentation rack with signal conditioning modules was acquired. This system has been connected to the sensors in the laboratory test loop to provide complete instrumentation channels of the types that exist in nuclear power plants. A photograph of this system is shown in Figure 8.1 along with a computer and other equipment used for the laboratory tests and data acquisition. Included in the photograph is the on-line monitoring system; also referred to as the calibration reduction system (CRS).

Laboratory Test Loop

A test loop simulating the primary coolant system of a PWR was constructed and used to test the software packages and algorithms developed in the project. A photograph and a drawing of the loop are shown in Figures 8.2 and 8.3 respectively. The loop is made of PVC piping with an inside diameter (I.D.) of 3" or 4" depending on which section of the loop the piping is used. The loop is filled with filtered tap water. The water is circulated in the loop by a 500 GPM centrifugal pump. There are three test sections in the loop, each with a number of temperature and pressure sensors. An air-cooled heat exchanger provides control of the water temperature in the loop. The water temperature can be controlled at a reasonably constant level at any plateau between room temperature and 120°F. The pressure in the loop can reach 200 psi and flow rates of as high as 15 feet per second are achievable in the pipes where the sensors are installed.

The sensors in the loop are connected to the Westinghouse 7300 instrumentation or other signal conversion and signal conditioning equipment. The loop can be operated continuously for many days to obtain long-term performance data under controlled and known conditions.

Contract With Host Utility

Negotiations were carried out with selected utilities to help with the in-plant validation of the technologies being developed in this project. As a result, the Duke Power Company was selected as the host utility. AMS made a presentation to Duke Power Company and wrote

Table 8.1

Listing of Sensors Installed in the Laboratory Test Loop

Tag #	Description	Calibration	Output	Manufacturer	Model	Type
TE-1-1	HVL TEMP	30-130°F	0-10 VDC	RdF	21205	RTD
TE-1-2	HVL TEMP	30-130°F	0-10 VDC	WEED	N9004D-2B	RTD
TE-1-3	ELBOW #1 DCHG TEMP	30-130°F	0-10 VDC	OMEGA	200Ω	RTD
TE-1-4	LVL TEMP	30-130°F	0-10 VDC	OMEGA	200Ω	RTD
TE-1-5	PUMP DCHG TEMP	30-130°F	0-10 VDC	OMEGA	200Ω	RTD
TE-2-1	HVL TEMP	30-130°F	1-5 VDC	WEED	N9004D-2B	RTD
TE-2-2	HX RETURN TEMP	30-130°F	1-5 VDC	RdF	21297	RTD
TE-3-2	LVL TEMP	200Ω DIN	OHMS	OMEGA	200Ω	RTD
TE-3-3	HVL TEMP	200Ω DIN	OHMS	OMEGA	200Ω	RTD
TE-3-4	HVL TEMP	100Ω DIN	OHMS	OMEGA	100Ω	RTD
TE-3-5	HVL TEMP	100Ω DIN	OHMS	OMEGA	100Ω	RTD
TE-3-6	ELBOW #2 DCHG TEMP	100Ω DIN	OHMS	OMEGA	100Ω	RTD
TC-1-1	HVL TEMP	32.6-250°F	0-10 VDC	OMEGA	TYPE J	TC
TC-1-2	LVL TEMP	32.6-250°F	0-10 VDC	OMEGA	TYPE J	TC
TC-1-3	ELBOW #2 DCHG TEMP	32.6-250°F	0-10 VDC	OMEGA	TYPE J	TC
TC-1-4	HVL TEMP	32.6-250°F	0-10 VDC	OMEGA	TYPE K	TC
TC-1-5	HVL TEMP	32.6-250°F	0-10 VDC	OMEGA	TYPE J	TC
TC-1-6	HX RETURN TEMP	32.6-250°F	0-10 VDC	OMEGA	TYPE J	TC
TC-1-7	RECIRC TEMP	32.6-250°F	0-10 VDC	OMEGA	TYPE J	TC
TC-0-1	HX AIR TEMP	32.6-250°F	1-5 VDC	OMEGA	TYPE J	TC
TC-0-2	HX INLET TEMP	32.6-250°F	1-5 VDC	OMEGA	TYPE J	TC
TC-0-3	HX OUTLET TEMP	32.6-250°F	1-5 VDC	OMEGA	TYPE J	TC
TC-S-1	HVL TEMP	30-130°F	1-5 VDC	OMEGA	TYPE E	TC
FT-1-1	ELBOW #1 D/P	5-0 PSI	0-10 VDC	BARTON	764	D/P
FT-1-2	ELBOW #1 D/P	0-5 PSI	0-10 VDC	FOXBORO	E13DM	D/P
FT-1-3	ELBOW #1 D/P	5-0 PSI	0-10 VDC	ROSEMOUNT	1153	D/P
FT-1-4	ELBOW #1 D/P	0-7.24 PSI	1-5 VDC	STATHAM	PD3200	D/P
FT-2-1	ELBOW #2 D/P	5-0 PSI	0-10 VDC	BARTON	764	D/P
FT-2-2	ELBOW #2 D/P	5-0 PSI	0-10 VDC	BARTON	764	D/P
FT-2-3	ELBOW #2 D/P	0-5 PSI	0-10 VDC	TOBAR	76DP	D/P
FT-2-4	ELBOW #2 D/P	0-10 PSI	1-5 VDC	STATHAM	PDH3200	D/P
FT-P-1	LVL FLOW	5-0 PSI	1-5 VDC	ROSEMOUNT	1153	D/P
FT-P-2	HX FLOW	0-5 PSI	1-5 VDC	ROSEMOUNT	3051C (Smart)	D/P
FT-P-3	MAIN LOOP FLOW	0-5 PSI	1-5 VDC	ROSEMOUNT	3051C (Smart)	D/P
FT-T-1	MAIN LOOP FLOW	0-500 GPM	0-5 VDC	M-C CONTROLS	E SERIES	PW
FT-3-1	PUMP D/P	0-15 PSI	1-5 VDC	ROSEMOUNT	3051C (Smart)	D/P
FT-3-2	PUMP D/P	0-15 PSI	0-10 VDC	ROSEMOUNT	1153	D/P
PT-1	LOOP PRESSURE	0-150 PSI	0-10 VDC	ROSEMOUNT	1152	PSR
PT-2	LOOP PRESSURE	0-150 PSI	0-10 VDC	FOXBORO	E11DM	PSR
PT-3	LOOP PRESSURE	0-390 PSI	1-5 VDC	FISCHER & PORTER	50EP1041	PSR

DCHG	=	Discharge	LVL	=	Low Velocity Leg
D/P	=	Differential Pressure	PSR	=	Pressure
DIN	=	German Standard for RTDs	PW	=	Paddlewheel
GPM	=	Gallons Per Minute	RECIRC	=	Recirculation
HVL	=	High Velocity Leg	TC	=	Thermocouple
HX	=	Heat Exchange	TEMP	=	Temperature

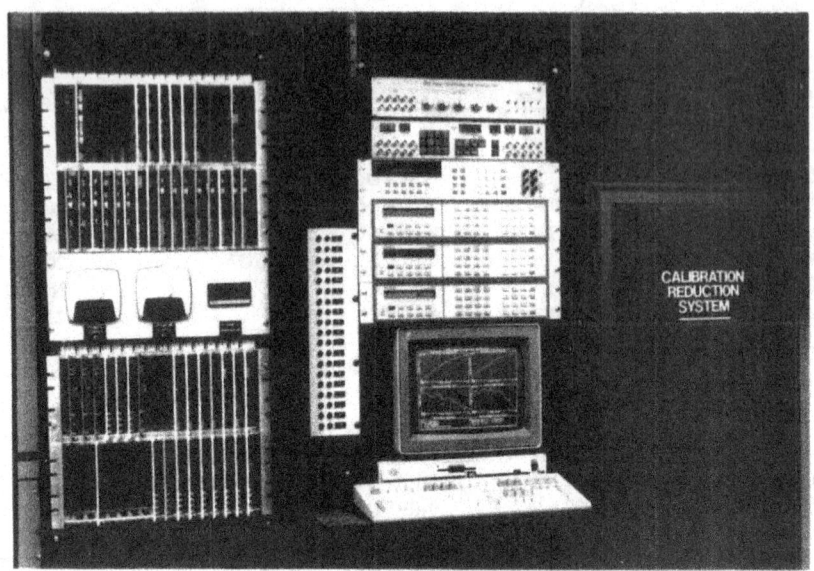

7300 System   Other Data Acquisition   CRS
Equipment

AMS-DWG BLK060A

Figure 8.1   Signal conditioning and signal conversion equipment used with
the sensors in the laboratory test loop

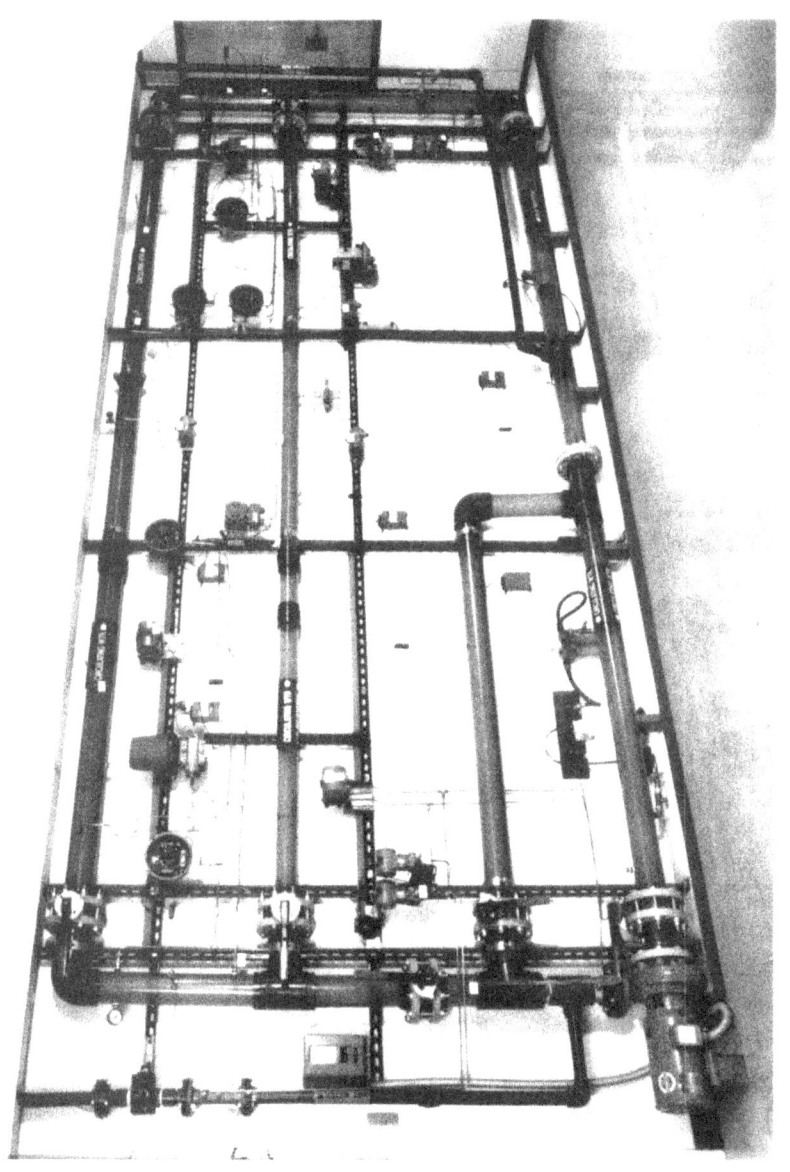

Figure 8.2   Laboratory test loop

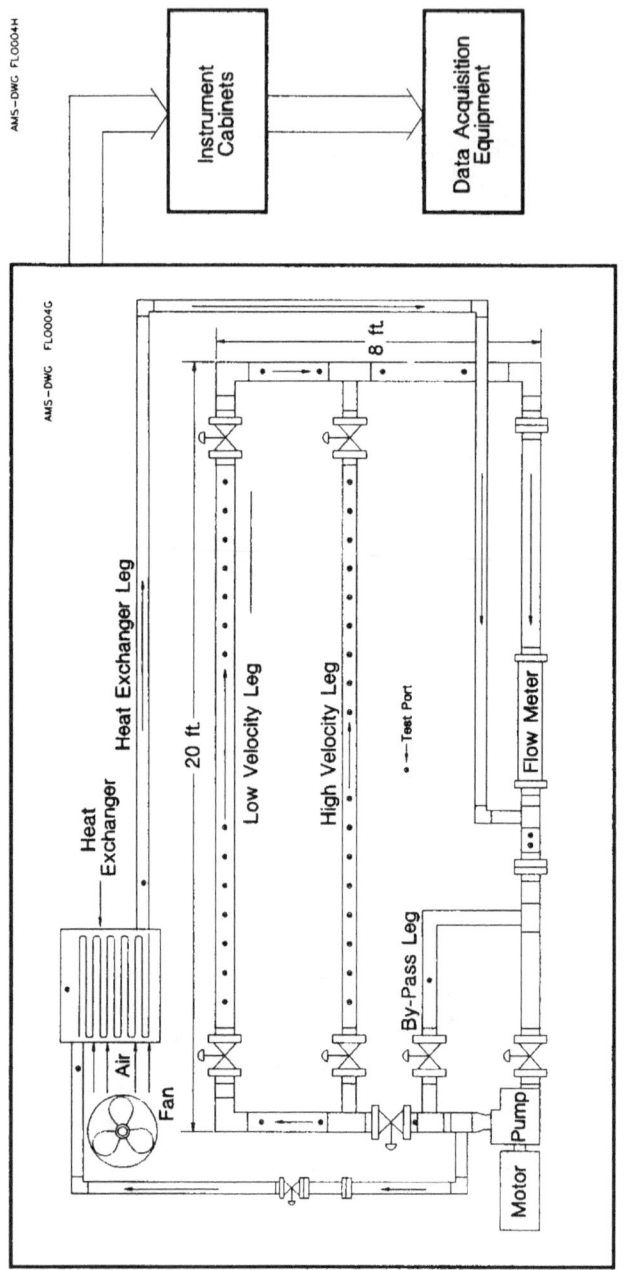

Figure 8.3 Test loop diagram

a formal proposal that resulted in a no-cost contract to perform a joint validation project on a long-term basis at the McGuire Nuclear Power Station Unit 2. McGuire Unit 2 is a four-loop Westinghouse PWR located in North Carolina. The unit went into commercial operation in March 1984.

Software Development

Through consulting agreements with some of the experts in this field, an assessment was made of previously developed software packages that could be acquired for the project. As a result, it was concluded that: 1) the data acquisition software packages already available at AMS are adequate for the project, and 2) it is best for AMS to write its own data analysis routines for this project.

The necessary algorithms were assembled during Phase I and successfully tested and implemented in an on-line monitoring system that was installed at the McGuire Nuclear Power Station. The system has been acquiring data from 170 live signals at McGuire since March 1992.

Search of LER Database

The NRC's Sequence Coding and Search System (SCSS) Licensee Event Report (LER) database, operated and maintained for the NRC by the Oak Ridge National Laboratory (ORNL) was searched to obtain historical data on performance of process instrumentation in nuclear power plants. The search was performed on all LERs with event dates from 1980 through September 1992. The results are summarized in Table 8.2 and Figures 8.4 and 8.5. The numbers in Table 8.2 and Figures 8.4 and 8.5 do not add up to 100% because of repetition of some problems in more than one category.

The LER results support our earlier statements that:

1. Only a small fraction of problems in process instrumentation channels are due to calibration drift. More specifically, 18 percent of instrument problems in the LER database are due to sensors and of this only 25 percent of problems are due to drift and other calibration problems.

2. Surveillance tests have frequently caused reactor trips. About 20 percent of all reactor trips reported in LERs in the last 12 years have been due to instrument testing.

3. Maintenance activities, including both acts of commission (e.g., shorting an instrument lead) and omission (e.g., inadequate preventive maintenance)

are major contributors to instrument failures.

It is important to point out in providing information from the LER database that as a result of a change in reporting requirement in 1984, problems such as the drift of a single instrument have not been reported in LERs since 1984. There are other databases such as: 1) the Nuclear Plant Reliability Data System (NPRDS) maintained by the Institute of Nuclear Power Operations (INPO), and 2) the Nuclear Operations Maintenance Information Service (NOMIS). The search of these and other databases will be performed in Phase II.

A summary of the observations discussed above is presented in Table 8.3.

8.2 Phase II Goals

A Phase II project has been awarded to AMS by the NRC and the work is currently underway. The Phase II project is a two-year effort to complete the research and development project started in Phase I. Following is a listing of the technical objectives of Phase II:

1. Quantify the accuracy of smart sensor technologies for fault detection and isolation in process instrumentation channels in nuclear power plants. This will help determine if the new technologies are as effective as the conventional techniques for verifying adequate sensor performance. The sensor faults of main interest in this project are calibration drift and response time degradation. The sensors of main interest in this project are pressure, level, and flow transmitters in the safety systems of pressurized and boiling water reactors.

   In addition to pressure, level, and flow transmitters, however, the Phase II project will include research on temperature, neutron flux, and other sensors in the McGuire Nuclear Power Plant.

2. Develop a database of sensor faults and the consequence of these faults on the static and dynamic responses of sensors. This effort will use on-line monitoring data to help determine the root cause of sensor drift and response time degradation.

3. Establish the requirements that may be considered by the NRC in allowing nuclear utilities to implement smart sensor technologies as a means of extending the frequency of calibration of process instrumentation channels, or eliminating some of the calibrations currently performed.

Table 8.2

**Summary of Search of LER Database for Reported
Problems with Nuclear Power Plant Instrumentation
(For the Period of 1980 to September 1992)**

Description	Total Number of LERs	Percent of Total
Total Number of LERs	38,048	100
Total Number of LERs Reporting Instrumentation Problems (including potential faults or failures)	21,073	55.4
Total Number of LERs Reporting Actual Instrumentation Faults or Failures	19,255	50.6
Total Number of LERs Reporting Actual Faults or Failures of Sensors (indicators, transmitters or primary sensing elements for pressure, level, flow or temperature)	3,641	9.6
Total Number of LERs Reporting Sensor Faults or Failures Due to Setpoint Drift or Other Calibration Problem	862	2.3
Total Number of LERs Reporting Plant Trips Due to Surveillance Tests Being Performed on Instrumentation (any instrumentation)	707	1.9
Total Number of LERs Reporting Actual Instrumentation Faults or Failures (any instrumentation) Due to Maintenance or Testing Activities	5,580	14.7

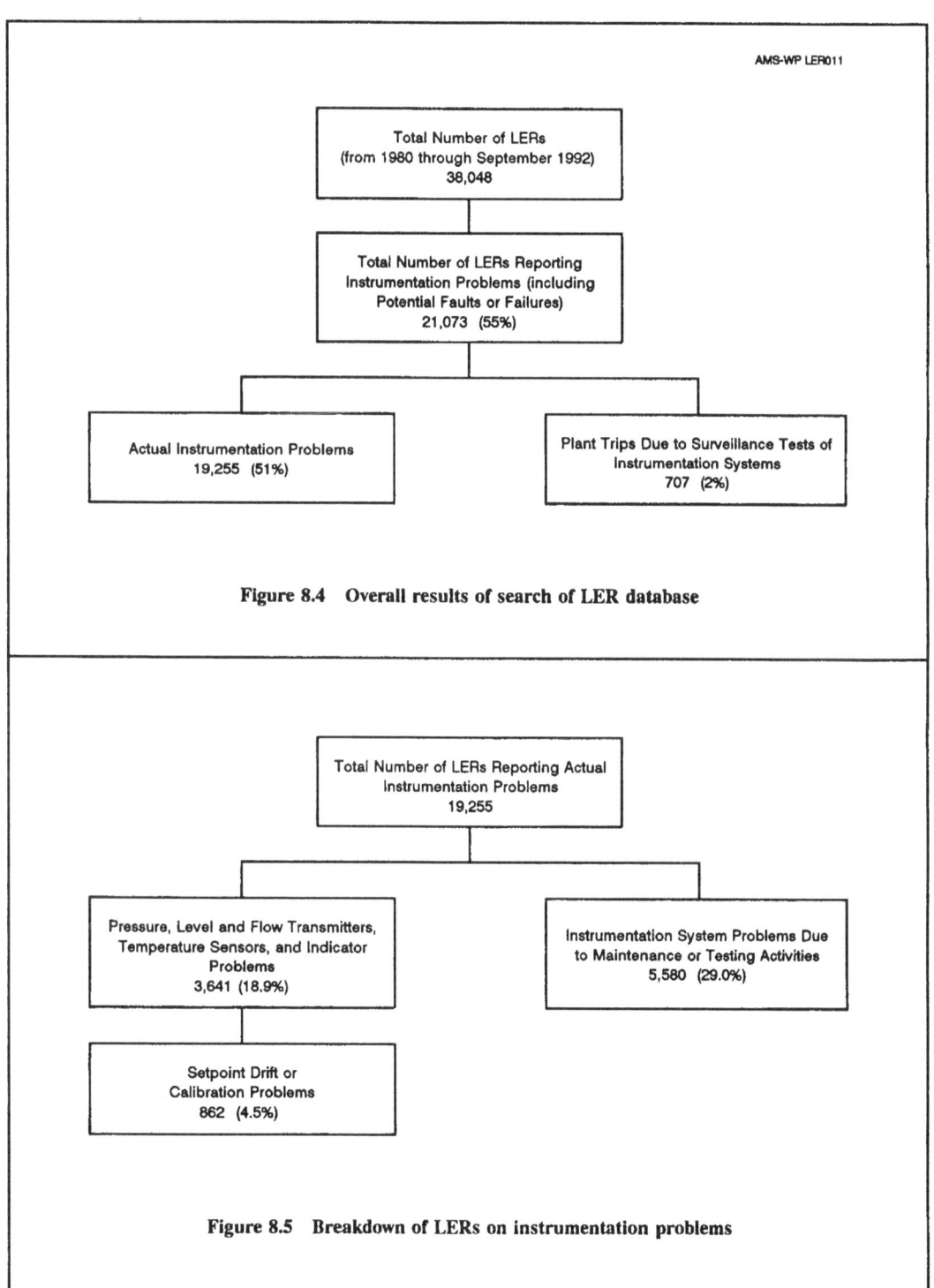

Total Number of LERs
(from 1980 through September 1992)
38,048

Total Number of LERs Reporting
Instrumentation Problems (including
Potential Faults or Failures)
21,073 (55%)

Actual Instrumentation Problems
19,255 (51%)

Plant Trips Due to Surveillance Tests of
Instrumentation Systems
707 (2%)

Figure 8.4   Overall results of search of LER database

Total Number of LERs Reporting Actual
Instrumentation Problems
19,255

Pressure, Level and Flow Transmitters,
Temperature Sensors, and Indicator
Problems
3,641 (18.9%)

Instrumentation System Problems Due
to Maintenance or Testing Activities
5,580 (29.0%)

Setpoint Drift or
Calibration Problems
862 (4.5%)

Figure 8.5   Breakdown of LERs on instrumentation problems

Table 8.3

Summary of Observations from the Search of LER Database

- 18% of instrument problems are due to sensors.

- 25% of sensor problems are due to drift and calibration.

- Instrument testing caused 707 plant trips. This is approximately 20 percent of all plant trips that have occurred in the 12 years of LER reporting period.

- Personnel error associated with maintenance testing is a frequent contributor to instrument system problems. About 30% of instrumentation problems are attributed to personnel error.

---

4. Develop an expert system for on-line verification of steady-state and transient performance of process sensors and the associated instrument channels in nuclear power plants.

5. Evaluate related developments by other organizations and use the information in the Phase II project.

6. Prepare Quality Assurance test procedures to be used in verification, validation, and implementation of the product of this project.

7. Develop and implement a plan to commercialize the results of the Phase II effort.

The technologies being evaluated or developed in this project can cover not only the process sensors, but also the signal conversion and signal conditioning equipment and other components of an instrument channel.

9. Development of On-Line Monitoring System

An on-line monitoring system generally consists of a data acquisition unit and a data analysis unit. Sometimes the two units are integrated and the data is analyzed in real time, and other times, the two units are used separately. We have used the second approach in this project until after the on-line monitoring techniques are successfully validated, at which time an integrated expert system will be developed and commercialized for use in nuclear power plants.

The data acquisition system developed in Phase I consists of a data acquisition cabinet (Figure 9.1) and an IBM compatible PC with three hard disk units; one removable and two fixed internal hard disks for redundant storage of a large volume of data. The system was delivered to the McGuire Nuclear Station in February 1992 and connected to 170 redundant and non-redundant signals from the safety-related and non-safety-related instrument channels in the primary and secondary systems of the plant.

The data acquisition cabinet contains the following modules which are controlled by the data acquisition computer:

- Four Hewlett-Packard (HP) Model 3488 switch-control (multiplexer) units. Each multiplexer contains 50 channels for a total of 200 channels. Additional channels can readily be added as needed.

- One HP Model 3457A digital multimeter. The multimeter is connected to the multiplexers and the computer via an IEEE 488 bus. Figure 9.2 shows a block diagram of the system.

- A signal isolator to prevent the on-line monitoring system from having any influence on the signals it is monitoring. Although the plant signals are isolated by the plant's own isolators, an isolation unit was installed in the data acquisition cabinet at the request of Duke Power Company to provide added protection. The plant signals pass through the isolator before they are fed to the HP multimeter in the data acquisition system. The isolator can be seen in the photograph of equipment wiring shown in Figure 9.3. The isolator provides about 5 gigohms (10^9 ohms) of isolation. Without this isolator, the 10 megohms (10^6 ohms) input impedance of the HP multimeter would have

been less than the McGuire plant requirement of 11 megohms.

- A stable DC power supply to check the calibration of the on-line monitoring system. The power supply voltage is measured by the data acquisition system at every scan and the results are stored and periodically examined to ensure that any drift in the data acquisition equipment is accounted for and corrected. Figure 9.4 shows a plot of the calibration signal sampled by the on-line monitoring system since the equipment was installed at McGuire in March 1992. It is clear that the equipment has not experienced a significant calibration drift in the nine months that it has been operating. The output of the DC power supply is periodically measured with a calibrated voltage standard to maintain traceability to the National Institute of Standards and Technology.

Once every 25 minutes the data acquisition system sequences through the 170 McGuire channels to collect DC data and store them on a hard disk. The scan of DC data requires approximately 15 minutes. The remaining ten minutes in the run is spent collecting AC data. This ten minute period is sufficient to collect 10 blocks of AC data on one channel. The 10 blocks of AC data are adequate for detecting gross degradation of response time in an instrument channel or sensor. Each block of AC data contains 1,024 points with a point sampled every 0.05 seconds. Of the 170 instrument channels monitored at McGuire, 110 channels have been selected for AC monitoring.

In summary, every 25 minutes onc DC data point is collected on each of the 170 channels being monitored for drift and one AC data set is collected on one of the 110 channels being monitored for response time degradation. The AC data acquisition requires about two days to complete all 110 channels during which time DC data points are still collected once every 25 minutes. The DC data is collected continuously, and the AC data are collected once a week. A high pass filter, an amplifier, and a low-pass filter are used to condition the signals for AC data collection (Figure 9.5). The high-pass filter is set at 0.02 Hz to remove the DC component of each signal so that the signal can be amplified for a better signal-to-noise characteristic. The amplifier is set at 20 db gain, and the low pass filter is set at 8 Hz for anti-aliasing and removal of extraneous noise.

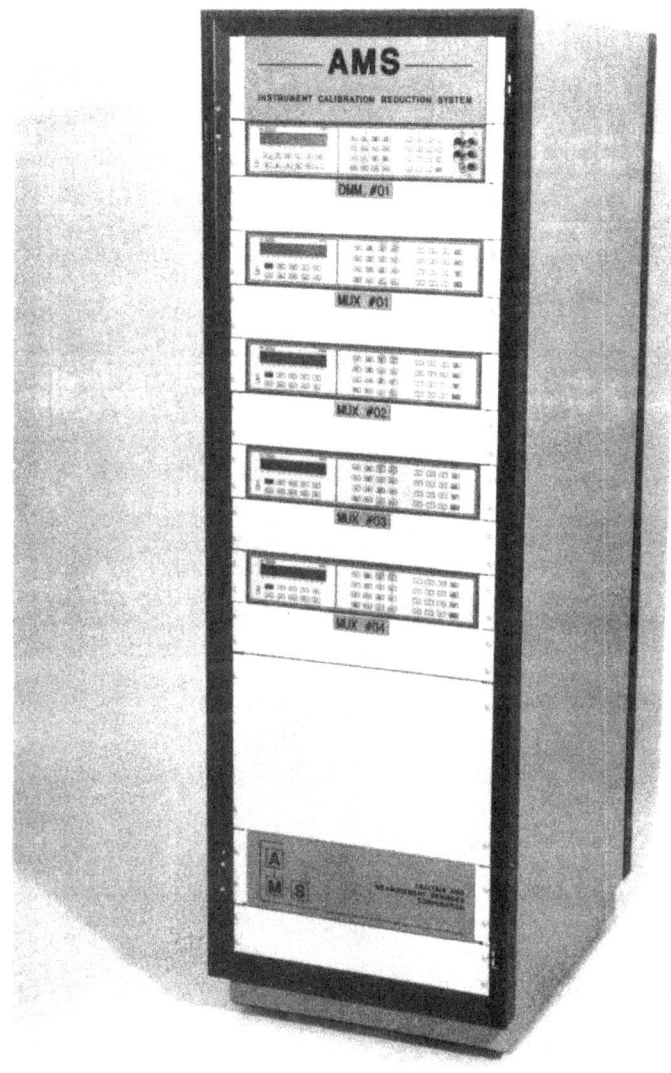

Figure 9.1 Data Acquisition Cabinet installed at McGuire for in-plant validation tests

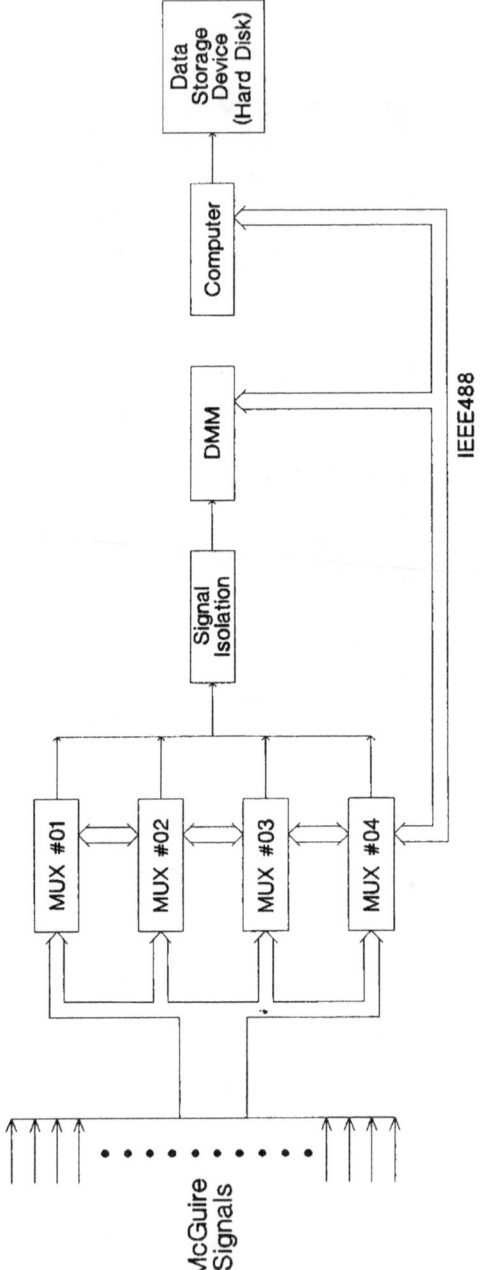

Figure 9.2 Block diagram of the On-Line Monitoring System

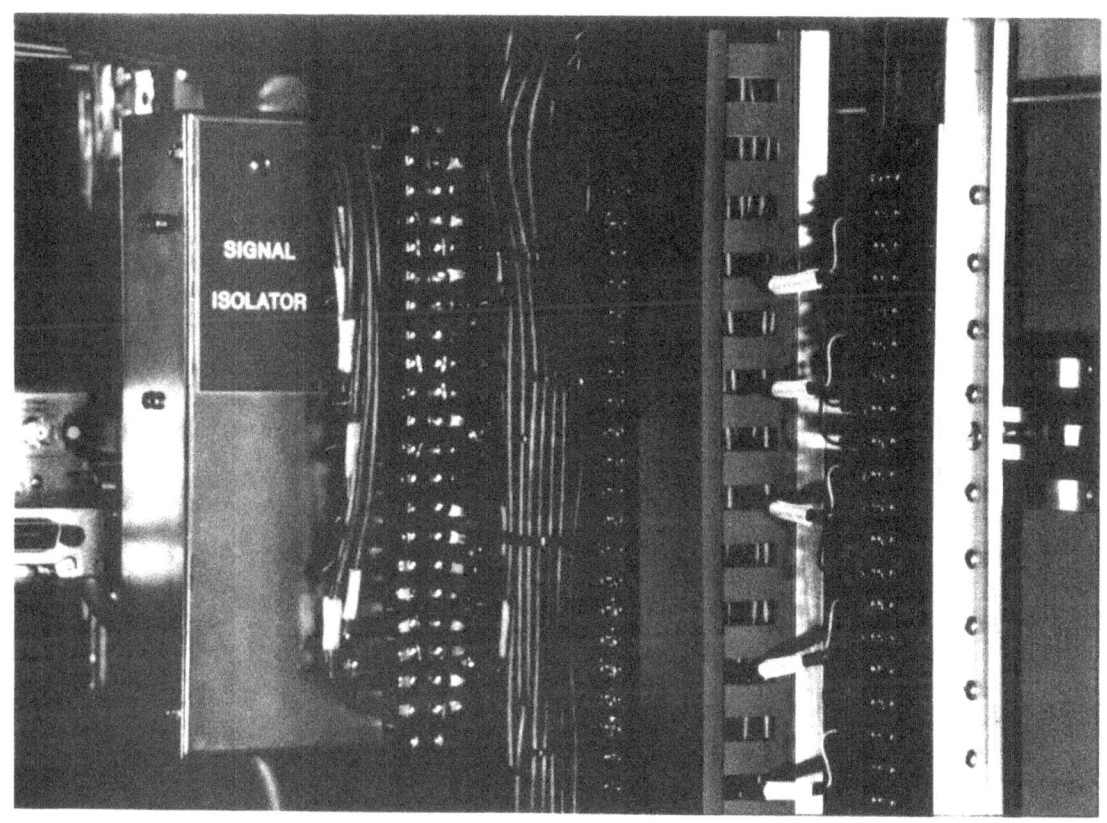

Figure 9.3   Wiring details of the On-Line Monitoring System

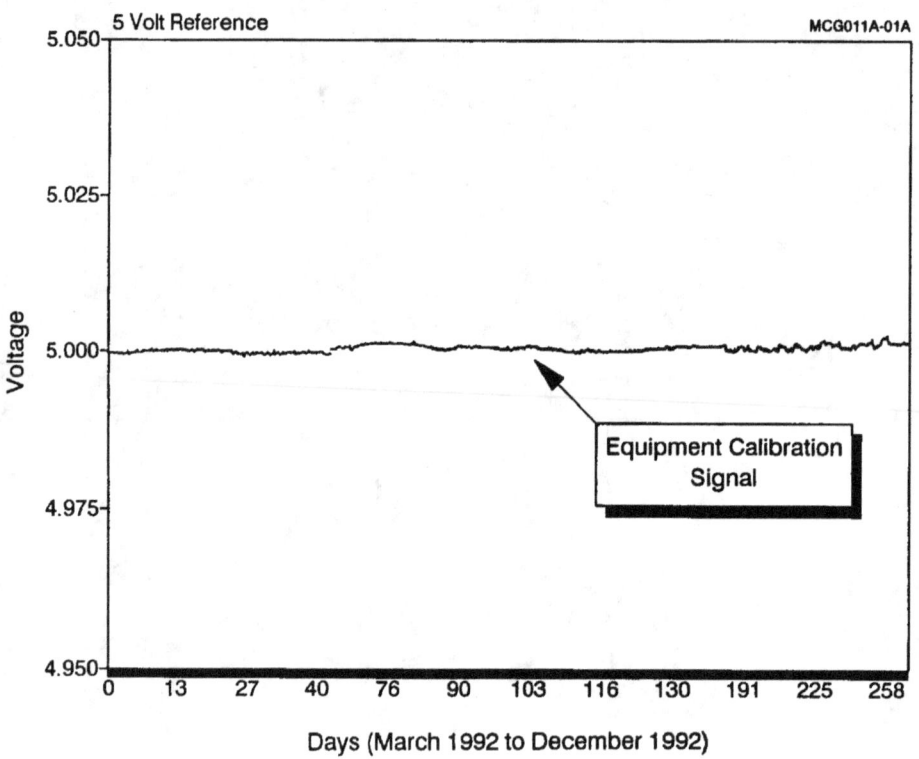

Figure 9.4   Calibration stability of Data Acquisition System

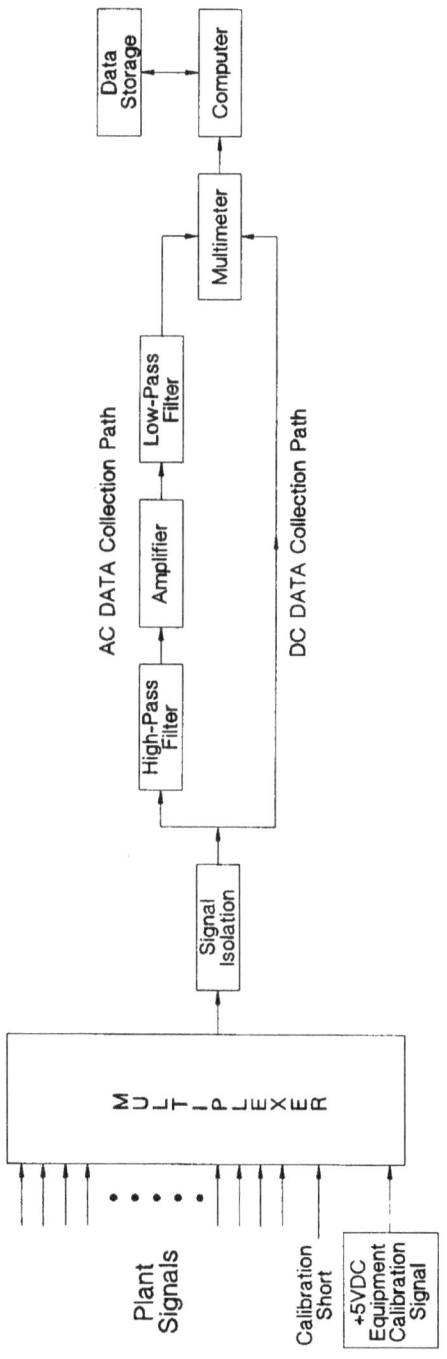

Figure 9.5 AC and DC data collection paths in the On-Line Monitoring System

Once every month the McGuire data disks are sent to AMS for analysis. The analysis is performed on a PC, and the results are plotted and visually examined for static and dynamic performance problems in the McGuire instrument channels. The analysis begins with a data preparation effort involving a limit checking and a transient removal process to minimize the effects of high frequency noise, spikes, and discontinuities in the data. Discontinuities occur in the data when there is a reactor trip during the data collection process; they also occur during the monthly surveillance tests while a channel is placed in test. Figure 9.6 shows an example of a McGuire data record before and after correction for discontinuities. This example has more than a normal number of gaps and discontinuities and is not typical of the data being monitored at the plant. Nevertheless, it was selected to show how even the worst type of data set can be successfully preprocessed and analyzed. The sudden jumps and discontinuities in the data correspond to those periods when the data collection had to be halted or the signal was not available for monitoring. This includes those occasions when the plant tripped or was shut down for maintenance.

It is apparent in the three graphs of Figure 9.6 that the spikes and discontinuities in the data can successfully be removed. Once the discontinuities are removed, the residual data are joined together to create a new data set that can be analyzed without difficulty. With this procedure, a gap may be observed in the corrected data indicating that the original data have had a step change due to a plant trip or another event (Figure 9.7). The gap does not cause a problem with the data analysis.

A plot of DC data records for three flow transmitters is shown in Figure 9.8. The data appear quite noisy. To overcome the noise problem, the sampled data are usually processed with an averaging window with a width of 10 to 100 times the sampling rate of the DC data. The width of the window is selected based on the type of signal and amount of noise on the data. The traces on the bottom plot of Figure 9.8 are the same data as the top plot with the noise removed by window averaging. It is apparent that the window averaging is effective in removing the high frequency components of the data which are not useful for drift analysis. It should be pointed out that the data are not put through any window averaging when they are sampled in the AC mode.

The window averaging has the same effect as a low-pass filter, meaning that it will increase the dynamic response results for the instrument being monitored. Another effect of window averaging is that it causes sudden signal changes to appear as transients. Again, this is not usually a problem in analyzing the DC data for detection of drift.

Once the data are corrected for discontinuities and digitally filtered by window averaging, they are analyzed to identify any drift in the instruments monitored. We have used only the simple techniques in the Phase I project for the analysis of the in-plant data. Basically, we perform an inconsistency test on the data and average the signals that are determined to be consistent. The resulting average is used as the best estimate of the process parameter and the deviation of each signal from the average is plotted and reviewed to identify any problems in the performance of the instrument. Figure 9.9 shows a block diagram of the current steps taken in analysis of a DC data record. Figure 9.10 shows a block diagram of the additional software modules including the AC analysis modules being developed and validated in Phase II.

The on-line monitoring system can be connected at any point in an instrument channel. Figure 9.11 shows a block diagram of typical components of a temperature channel. If the on-line monitoring system is connected at the end of the channel as shown in the figure, then the drift and response time degradation of all the components in the channel can be determined in a single test. However, the main interest of the nuclear power industry is in remote testing of the sensor more so than the other components of the instrument channel. This is especially true in the case of pressure, level, and flow transmitters. For this reason, we will continue to concentrate our in-plant validation efforts in Phase II on the pressure, level, and flow transmitters at McGuire. In the meantime, we will monitor other in-plant sensors to: 1) provide a database of diverse signals to be used as a learning tool to study the behavior and relationship of redundant and non-redundant signals, and 2) provide an adequate number of signals for the analytical redundancy models to account for common mode drift. With these points in mind, the signals listed in Table 9.1 were selected for the in-plant validation tests at McGuire.

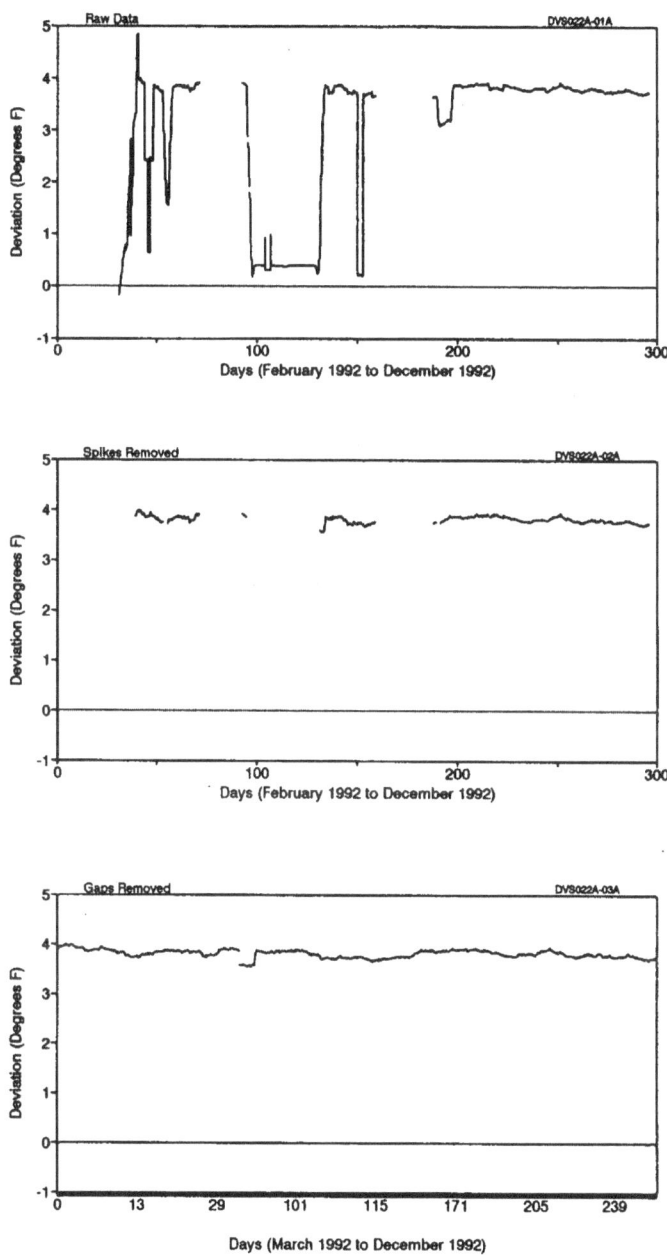

Figure 9.6   Example of discontinuities, gaps, and spikes in a raw data record and
how they are removed to prepare the data for drift analysis

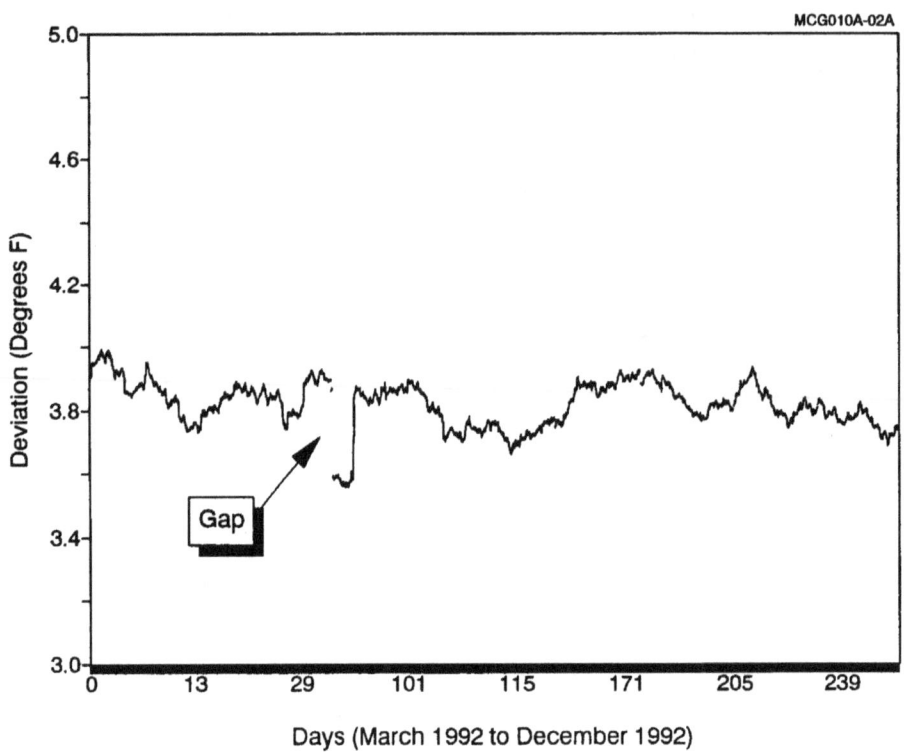

Figure 9.7   Illustration of gap in a corrected data record

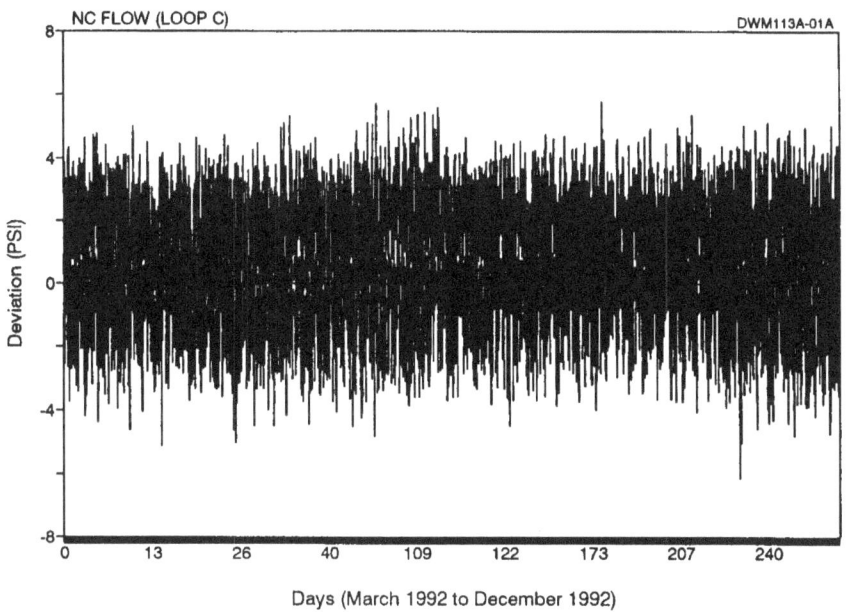

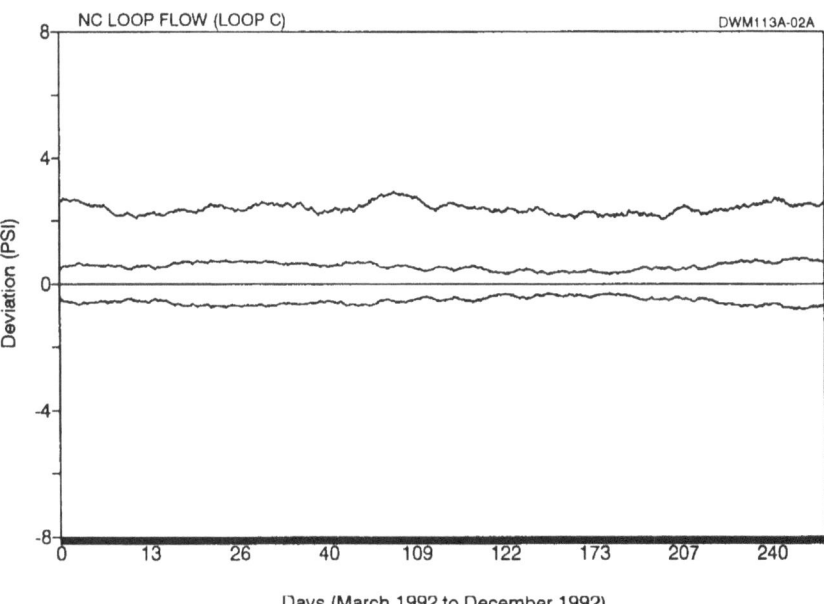

Figure 9.8   Signals before and after window averaging

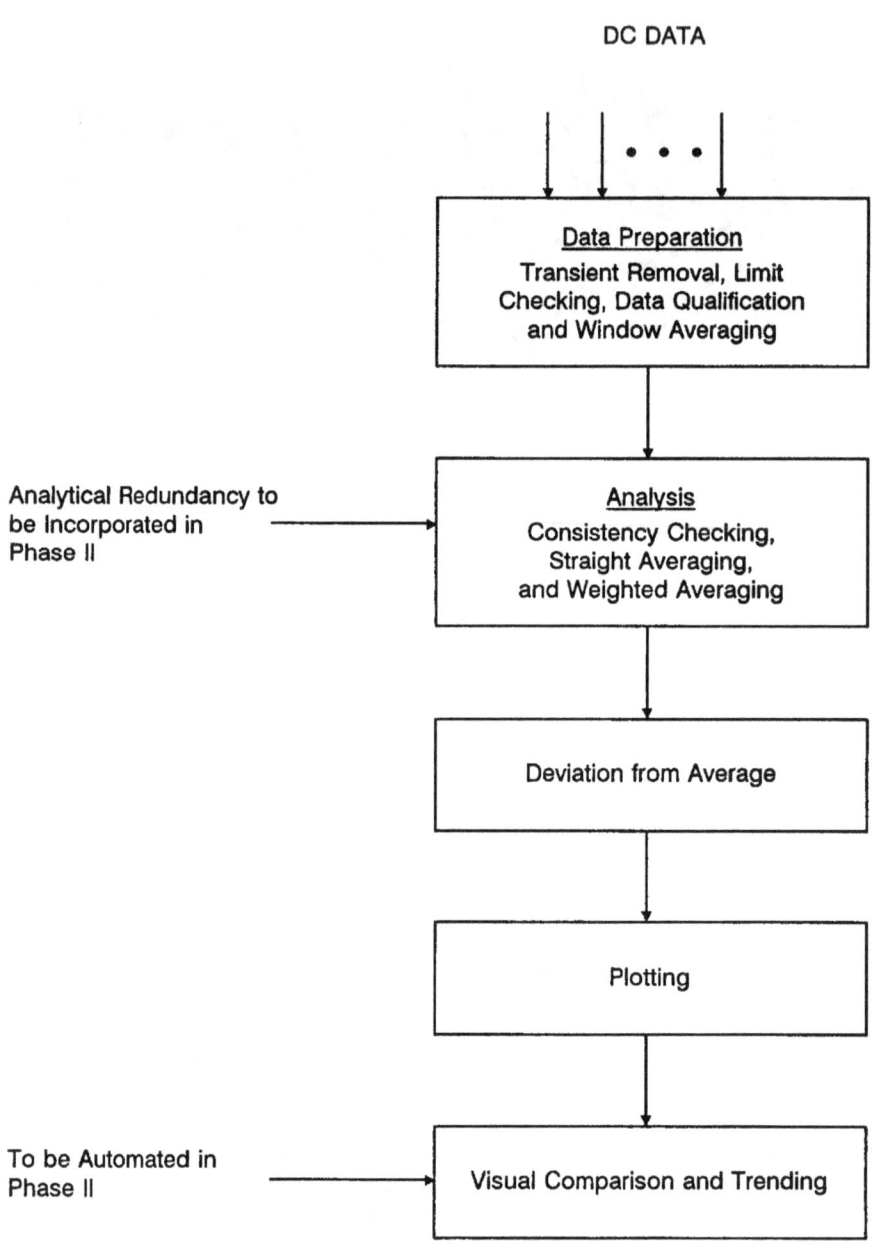

Figure 9.9   Block diagram of steps taken in Phase I for analysis of DC data

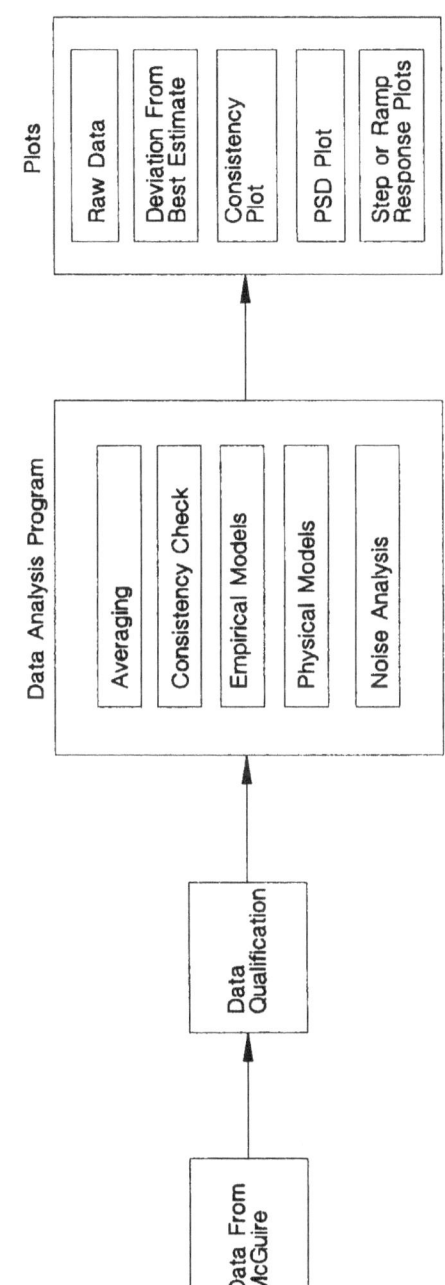

Figure 9.10 Block diagram of signal analysis modules to be incorporated in the final test equipment

RTD Channel

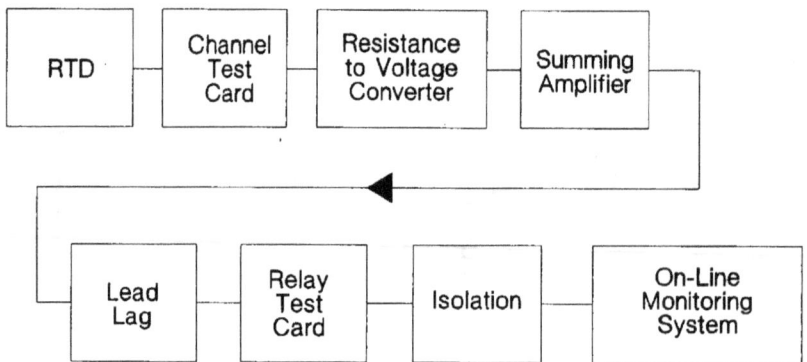

**Figure 9.11 Typical components of an instrument channel tested together
by the on-line monitoring system**

Table 9.1

Listing of Signals Monitored at the McGuire Nuclear Power Station Unit 2

Item	Tag #	Description	Operating Range	Units	Sensor Make/Model
		Auxiliary Feedwater Flow			
1	CAFT5090	STM GEN A	0-300	GPM	Rosemount 1151
2	CAFT5100	STM GEN B	0-300	GPM	Rosemount 1151
3	CAFT5110	STM GEN C	0-300	GPM	Rosemount 1151
4	CAFT5120	STM GEN D	0-300	GPM	Rosemount 1151
		Feedwater Flow			
5	CFFT5000	STM GEN A #1	0-118	%	Rosemount 3051C
6	CFFT5010	STM GEN A #2	0-118	%	Rosemount 3051C
7	CFFT5020	STM GEN B #1	0-118	%	Rosemount 3051C
8	CFFT5030	STM GEN B #2	0-118	%	Rosemount 3051C
9	CFFT5040	STM GEN C #1	0-118	%	Rosemount 3051C
10	CFFT5050	STM GEN C #2	0-118	%	Rosemount 3051C
11	CFFT5060	STM GEN D #1	0-118	%	Rosemount 3051C
12	CFFT5070	STM GEN D #2	0-118	%	Rosemount 3051C
		Steam Generator Level			
13	CFLT5490	NR LEVEL STM GEN A #4	0-100	%	Barton 764
14	CFLT5500	NR LEVEL STM GEN A #3	0-100	%	Barton 764
15	CFLT5510	NR LEVEL STM GEN A #2	0-100	%	Barton 764
16	CFLT6000	NR LEVEL STM GEN A #1	0-100	%	Barton 764
17	CFLT5610	WR LEVEL STM GEN A	0-100	%	Rosemount 1153
18	CFLT5520	NR LEVEL STM GEN B #4	0-100	%	Barton 764
19	CFLT5530	NR LEVEL STM GEN B #3	0-100	%	Barton 764
20	CFLT5540	NR LEVEL STM GEN B #1	0-100	%	Barton 764
21	CFLT6010	NR LEVEL STM GEN B #2	0-100	%	Barton 764
22	CFLT5620	WR LEVEL STM GEN B	0-100	%	Rosemount 1153
23	CFLT5550	NR LEVEL STM GEN C #4	0-100	%	Barton 764
24	CFLT5560	NR LEVEL STM GEN C #3	0-100	%	Barton 764
25	CFLT6020	NR LEVEL STM GEN C #2	0-100	%	Barton 764
26	CFLT5570	NR LEVEL STM GEN C #1	0-100	%	Barton 764
27	CFLT5630	WR LEVEL STM GEN C	0-100	%	Rosemount 1153
28	CFLT5580	NR LEVEL STM GEN D #4	0-100	%	Barton 764
29	CFLT5590	NR LEVEL STM GEN D #3	0-100	%	Barton 764
30	CFLT5600	NR LEVEL STM GEN D #2	0-100	%	Barton 764
31	CFLT6030	NR LEVEL STM GEN D #1	0-100	%	Barton 764
32	CFLT5640	WR LEVEL STM GEN D	0-100	%	Rosemount 1153
		Reactor Coolant Flow			
33	NCFT5000	NC LOOP FLOW A #1	0-120	%	Barton 764
34	NCFT5010	NC LOOP FLOW A #2	0-120	%	Barton 764
35	NCFT5020	NC LOOP FLOW A #3	0-120	%	Barton 764
36	NCFT5030	NC LOOP FLOW B #1	0-120	%	Barton 764
37	NCFT5040	NC LOOP FLOW B #2	0-120	%	Barton 764
38	NCFT5050	NC LOOP FLOW B #3	0-120	%	Barton 764
39	NCFT5060	NC LOOP FLOW C #1	0-120	%	Barton 764
40	NCFT5070	NC LOOP FLOW C #2	0-120	%	Barton 764
41	NCFT5080	NC LOOP FLOW C #3	0-120	%	Barton 764
42	NCFT5090	NC LOOP FLOW D #1	0-120	%	Barton 764
43	NCFT5100	NC LOOP FLOW D #2	0-120	%	Barton 764
44	NCFT5110	NC LOOP FLOW D #3	0-120	%	Barton 764
		Pressurizer Level			
45	NCLT5150	PROTECTION #1	0-100	%	Barton 764
46	NCLT5160	PROTECTION #2	0-100	%	Barton 764
47	NCLT5170	PROTECTION #3	0-100	%	Barton 764

Continued on Next Page

Table 9.1 (Continued)

Item	Tag #	Description	Operating Range	Units	Sensor Make/Model
		RVLIS			
48	NCLT6630	UPPER TRAIN A	60-120	%	Barton 752
49	NCLT6640	LOWER TRAIN A	0-70	%	Barton 752
50	NCLT6650	D/P TRAIN A	0-120	%	Barton 752
51	NCLT6660	UPPER TRAIN B	60-120	%	Barton 752
52	NCLT6670	LOWER TRAIN B	0-70	%	Barton 752
53	NCLT6680	D/P TRAIN B	0-120	%	Barton 752
		Wide Range Pressure			
54	NCPT5120	HOT LEG LOOP D	0-3000	PSI	Rosemount 1153
55	NCPT5140	HOT LEG LOOP C	0-3000	PSI	Barton 763
		Pressurizer Pressure			
56	NCPT5150	PROTECTION #2	1700-2500	PSI	Barton 763
57	NCPT5160	PROTECTION #1	1700-2500	PSI	Barton 763
58	NCPT5170	PROTECTION #3	1700-2500	PSI	Barton 763
59	NCPT5171	PROTECTION #4	1700-2500	PSI	Barton 763
		Containment Pressure			
60	NSPT5040	PROTECTION #4	-5-20	PSI	Barton 386A
61	NSPT5050	PROTECTION #3	-5-20	PSI	Barton 386A
62	NSPT5060	PROTECTION #2	-5-20	PSI	Barton 386A
		Steam Flow			
63	SMFT5000	STM GEN A #1	0-4.542	MPPH	Barton 764
64	SMFT5010	STM GEN A #2	0-4.542	MPPH	Barton 764
65	SMFT5020	STM GEN B #2	0-4.542	MPPH	Barton 764
66	SMFT5030	STM GEN B #1	0-4.542	MPPH	Barton 764
67	SMFT5040	STM GEN C #1	0-4.542	MPPH	Barton 764
68	SMFT5050	STM GEN C #2	0-4.542	MPPH	Barton 764
69	SMFT5060	STM GEN D #2	0-4.542	MPPH	Barton 764
70	SMFT5070	STM GEN D #1	0-4.542	MPPH	Barton 764
		Steam Pressure			
71	SMPT5080	STM GEN A #1	0-1300	PSI	Rosemount 1153
72	SMPT5090	STM GEN A #2	0-1300	PSI	Rosemount 1153
73	SMPT5100	STM GEN A #4	0-1300	PSI	Rosemount 1153
74	SMPT5110	STM GEN B #1	0-1300	PSI	Rosemount 1153
75	SMPT5120	STM GEN B #2	0-1300	PSI	Rosemount 1153
76	SMPT5130	STM GEN B #3	0-1300	PSI	Rosemount 1153
77	SMPT5140	STM GEN C #1	0-1300	PSI	Rosemount 1153
78	SMPT5150	STM GEN C #2	0-1300	PSI	Rosemount 1153
79	SMPT5160	STM GEN C #3	0-1300	PSI	Rosemount 1153
80	SMPT5170	STM GEN D #1	0-1300	PSI	Rosemount 1153
81	SMPT5180	STM GEN D #2	0-1300	PSI	Rosemount 1153
82	SMPT5190	STM GEN D #4	0-1300	PSI	Rosemount 1153
		Turbine Impulse Pressure			
83	SMPT5210	CHAMBER PRESS #1	0-800	PSI	Rosemount 1153
84	SMPT5220	CHAMBER PRESS #2	0-800	PSI	Rosemount 1153
		Reactor Coolant Temperature			
85	NCRD8120	NR T-HOT A H1	530-650	Deg F	RdF RTD
86	NCRD8130	NR T-HOT A H2	530-650	Deg F	RdF RTD
87	NCRD8140	NR T-HOT A H3	530-650	Deg F	RdF RTD
88	NCRD5900	WR T-HOT A	0-700	Deg F	Conax RTD
89	NCRD8160	NR T-COLD A C1	510-630	Deg F	RdF RTD
90	NCRD5860	WR T-COLD A	0-700	Deg F	Conax RTD
91	NCRD8170	NR T-HOT B H1	530-650	Deg F	RdF RTD
92	NCRD8180	NR T-HOT B H2	530-650	Deg F	RdF RTD

Continued on Next Page

Table 9.1 (Continued)

Item	Tag #	Description	Operating Range	Units	Sensor Make/Model
		Reactor Coolant Temperature (Continued)			
93	NCRD8190	NR T-HOT B H3	530-650	Deg F	RdF RTD
94	NCRD5920	WR T-HOT B	0-700	Deg F	Conax RTD
95	NCRD8210	NR T-COLD B	510-630	Deg F	RdF RTD
96	NCRD5880	WR T-COLD B	0-700	Deg F	Conax RTD
97	NCRD8220	NR T-HOT C H1	530-650	Deg F	RdF RTD
98	NCRD8230	NR T-HOT C H2	530-650	Deg F	RdF RTD
99	NCRD8240	NR T-HOT C H3	530-650	Deg F	RdF RTD
		Primary Coolant Temperature			
100	NCRD5850	WR T-HOT C	0-700	Deg F	Conax RTD
101	NCRD8260	NR T-COLD C	510-630	Deg F	RdF RTD
102	NCRD5910	WR T-COLD C	0-700	Deg F	Conax RTD
103	NCRD8270	NR T-HOT D H1	530-650	Deg F	RdF RTD
104	NCRD8280	NR T-HOT D H2	530-650	Deg F	RdF RTD
105	NCRD8290	NR T-HOT D H3	530-650	Deg F	RdF RTD
106	NCRD5870	WR T-HOT D	0-700	Deg F	Conax RTD
107	NCRD8310	NR T-COLD D	510-630	Deg F	RdF RTD
108	NCRD5930	WR T-COLD D	0-700	Deg F	Conax RTD
		Loop ΔT			
109	A D/T	LOOP A D/T	0-150	%PU	7300
110	B D/T	LOOP B D/T	0-150	%PU	7300
111	C D/T	LOOP C D/T	0-150	%PU	7300
112	D D/T	LOOP D D/T	0-150	%PU	7300
		Loop T Average			
113	A TAVG	LOOP A TAVG	530-630	Deg F	7300
114	B TAVG	LOOP B TAVG	530-630	Deg F	7300
115	C TAVG	LOOP C TAVG	530-630	Deg F	7300
116	D TAVG	LOOP D TAVG	530-630	Deg F	7300
		Neutron Detectors			
117	N41 UL	POWER RANGE UPPER LEVEL Q4	0-120	%PU	N/A
118	N42 UL	POWER RANGE UPPER LEVEL Q2	0-120	%PU	N/A
119	N43 UL	POWER RANGE UPPER LEVEL Q1	0-120	%PU	N/A
120	N44 UL	POWER RANGE UPPER LEVEL Q3	0-120	%PU	N/A
121	N41 LL	POWER RANGE LOWER LEVEL Q4	0-120	%PU	N/A
122	N42 LL	POWER RANGE LOWER LEVEL Q2	0-120	%PU	N/A
123	N43 LL	POWER RANGE LOWER LEVEL Q1	0-120	%PU	N/A
124	N44 LL	POWER RANGE LOWER LEVEL Q3	0-120	%PU	N/A
125	N41 AVG	POWER RANGE AVG LEVEL Q4	0-120	%PU	N/A
126	N42 AVG	POWER RANGE AVG LEVEL Q2	0-120	%PU	N/A
127	N43 AVG	POWER RANGE AVG LEVEL Q1	0-120	%PU	N/A
128	N44 AVG	POWER RANGE AVG LEVEL Q3	0-120	%PU	N/A
		Incore Thermocouple			
129	B05	TRAIN A	32-2300	Deg F	N/A
130	B13	TRAIN A	32-2300	Deg F	N/A
131	C08	TRAIN A	32-2300	Deg F	N/A
132	D03	TRAIN A	32-2300	Deg F	N/A
133	D11	TRAIN A	32-2300	Deg F	N/A
134	F05	TRAIN A	32-2300	Deg F	N/A
135	F09	TRAIN A	32-2300	Deg F	N/A
136	F15	TRAIN A	32-2300	Deg F	N/A
137	G02	TRAIN A	32-2300	Deg F	N/A
138	G12	TRAIN A	32-2300	Deg F	N/A
139	J10	TRAIN A	32-2300	Deg F	N/A
140	K01	TRAIN A	32-2300	Deg F	N/A

Continued on Next Page

Table 9.1 (Continued)

Item	Tag #	Description	Operating Range	Units	Sensor Make/Model
		Incore Thermocouple (Continued)			
141	K05	TRAIN A	32-2300	Deg F	N/A
142	K13	TRAIN A	32-2300	Deg F	N/A
143	L08	TRAIN A	32-2300	Deg F	N/A
144	M11	TRAIN A	32-2300	Deg F	N/A
145	M13	TRAIN A	32-2300	Deg F	N/A
146	N02	TRAIN A	32-2300	Deg F	N/A
147	P05	TRAIN A	32-2300	Deg F	N/A
148	P09	TRAIN A	32-2300	Deg F	N/A
149	A06	TRAIN B	32-2300	Deg F	N/A
150	B11	TRAIN B	32-2300	Deg F	N/A
151	C04	TRAIN B	32-2300	Deg F	N/A
152	D07	TRAIN B	32-2300	Deg F	N/A
153	D13	TRAIN B	32-2300	Deg F	N/A
154	E02	TRAIN B	32-2300	Deg F	N/A
155	E10	TRAIN B	32-2300	Deg F	N/A
156	G04	TRAIN B	32-2300	Deg F	N/A
157	G14	TRAIN B	32-2300	Deg F	N/A
158	H11	TRAIN B	32-2300	Deg F	N/A
159	J02	TRAIN B	32-2300	Deg F	N/A
160	J08	TRAIN B	32-2300	Deg F	N/A
161	K15	TRAIN B	32-2300	Deg F	N/A
162	L02	TRAIN B	32-2300	Deg F	N/A
163	L12	TRAIN B	32-2300	Deg F	N/A
164	N04	TRAIN B	32-2300	Deg F	N/A
165	N06	TRAIN B	32-2300	Deg F	N/A
166	N10	TRAIN B	32-2300	Deg F	N/A
167	N14	TRAIN B	32-2300	Deg F	N/A
168	R08	TRAIN B	32-2300	Deg F	N/A
		Miscellaneous			
169	SHORT	LOW THERMAL SHORT	N/A	VDC	N/A
170	REF	5 VOLT REFERENCE	N/A	VDC	N/A

Deg F	=	Degree Fahrenheit	Q1 to Q4	=	Core Quadrant 1 to 4
D/P	=	Differential Pressure	RVLIS	=	Reactor Vessel Level Indicating System
GPM	=	Gallons Per Minute	STM GEN	=	Steam Generator
MPPH	=	Million Pounds Per Hour	T-COLD	=	Cold Leg Temperature
NC	=	Nuclear Coolant	T-HOT	=	Hot Leg Temperature
NR	=	Narrow Range	WR	=	Wide Range
PSI	=	Pounds per Square Inch			

10. Laboratory Test Results

The feasibility of smart sensor technologies was studied through a series of experiments performed on representative temperature and pressure sensors and associated signal conditioning equipment installed in a laboratory test loop. Figure 10.1 shows the type and location of the sensors as installed and tested in the loop. The laboratory test equipment and the details of the development of the loop were described earlier. The results of the laboratory tests are summarized in this chapter.

An empirical model of the laboratory loop was developed based on Equation 5.7 discussed in Chapter 5. The model was then validated with data from the sensors in the loop. Representative results are shown in Figure 10.2 in terms of time history plots for two flow transmitters. Each plot shows the model prediction and the corresponding measured data. It is apparent that the model prediction agrees well with the measured data indicating the validity of the model. As shown in Figure 10.2, the signal level was changed several times during the tests to help identify the parameters of the model. The same type of comparison for a physical model based on Equation 5.12 of Chapter 5 is shown in Figure 10.3. This data represents a transient temperature signal induced by a cool down of the test loop. The difference between the model prediction and the actual temperature is also shown. The differences throughout the data are small (less than 1 percent) indicating the validity of the physical model. Another example of transient temperature data is shown in Figure 10.4 for four thermocouples tested in the loop from start up through steady-state conditions. The model prediction is not shown in this figure but the actual loop temperature as measured with a calibrated RTD is shown. Following the initial transients during the loop start up, the thermocouples track the actual process temperature to within about 1°F as shown in the deviation plot of Figure 10.4. The deviation plot represents the difference between each signal and the average of all signals in the group. Similar data for step increases in the loop flow are shown in Figure 10.5 for three flow transmitters. The deviation plot in this figure shows that the three transmitters track each other very well. The significance of these plots and other data shown in this chapter is that they provide confirmation that the tests, data acquisition, and data analysis algorithms used in this project are performing as expected, and are providing a means to gain experience with the behavior of various sensors under various test conditions.

The loop data were also used to test the consistency checking and averaging algorithms. Figure 10.6 shows signals from four pressure transmitters measuring the same differential pressure across an elbow in the test loop. As shown in the figure, one of the four signals is manipulated to cause it to deviate from the rest of the signals. At about 30 minutes into the data, as the sensor deviation begins to exceed a preset value, the signal is determined to be inconsistent and is marked by a "*" placed on the affected portion of the trace. The consistency checking is performed on a continuous basis and the results are continuously updated. All redundant signals are included in the average while they are consistent with the other signals. Similar data to that of Figure 10.6 are shown in Figure 10.7 for five temperature sensors.

Figure 10.8 shows steady-state traces for three flow transmitters, with a bias intentionally induced in FT-1-3 to show how the bias manifests itself in the deviation plot. Similar data are shown in Figure 10.9 for three signals whose levels are changed in a step fashion throughout the experiment. It is clear that the sensor deviations are not affected by the step changes induced in the level of the signals. A pair of plots for another group of flow transmitters is shown in Figure 10.10 with one of the signals intentionally biased and another one manipulated during the tests to exhibit a drift. Note that the drift in FT-1-1 causes all three signals to exhibit drift behavior in the deviation plot. This is a case where pairwise comparison should be used to reveal which signal is actually drifting. Similar plots to that of Figure 10.10 are shown in Figures 10.11 and 10.12 for several temperature sensors. This is followed by a pair of plots for three differential pressure transmitters in Figure 10.13 in which two transmitters are adjusted to have the same steady-state output and one is intentionally drifted. Note the sudden shift in the deviation traces at about 600 minutes into the data. This occurs because FT-1-1 is excluded from the average at this point in time when its deviation exceeds a preset criterion. This causes all three signals to exhibit a step change in the deviation plot.

Figure 10.14 shows five temperature signals with one erratic signal in the group. The erratic signal is due to a noisy signal conditioning module (the signal conditioning module in this case was an NRA Card). Note that the erratic behavior of this signal affects the deviation plots of all signals in the group.

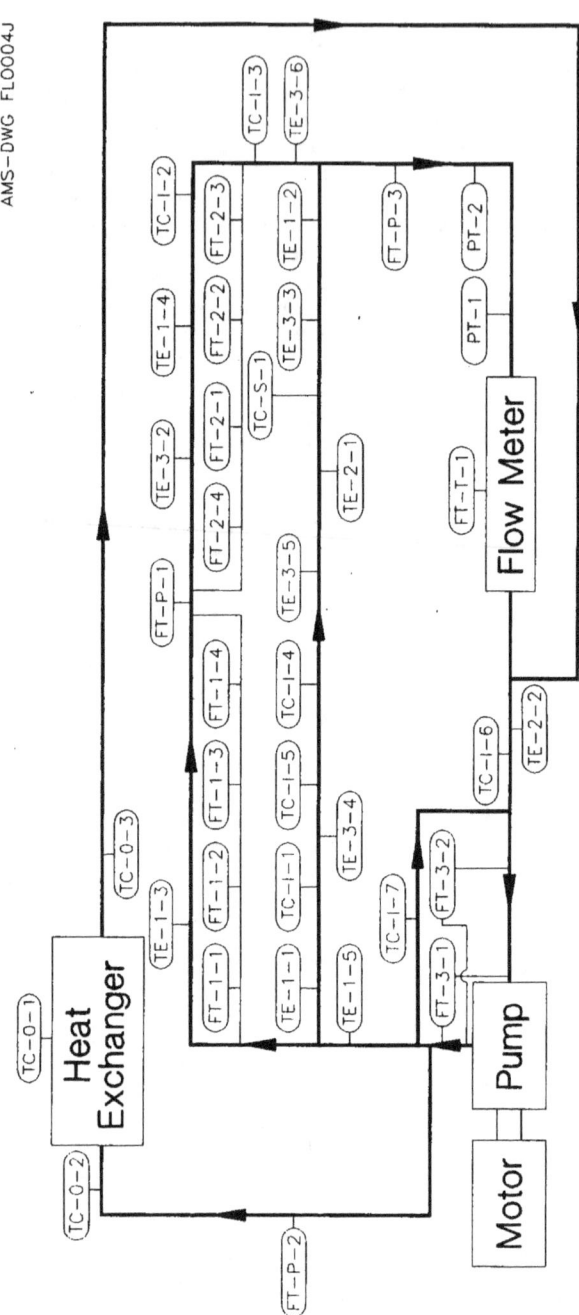

AMS—DWG FL0004J

Figure 10.1   Location, type and tag numbers of sensors installed in the laboratory test loop

- 100 -

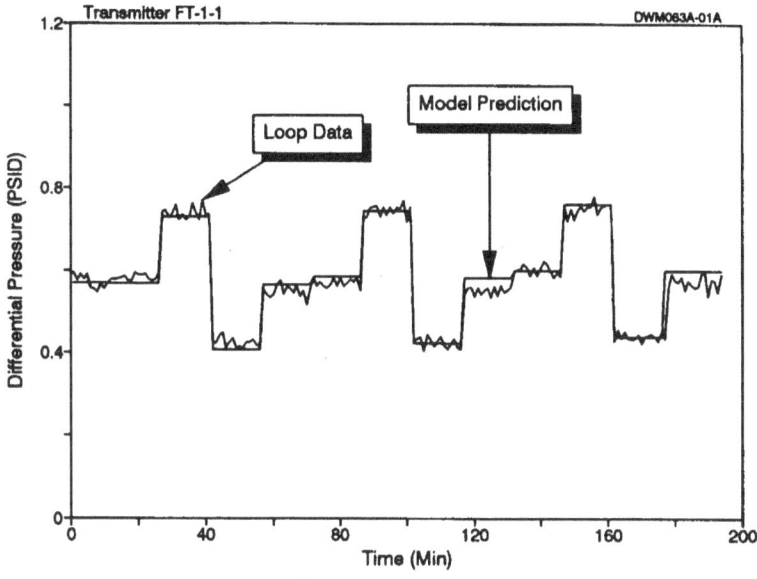

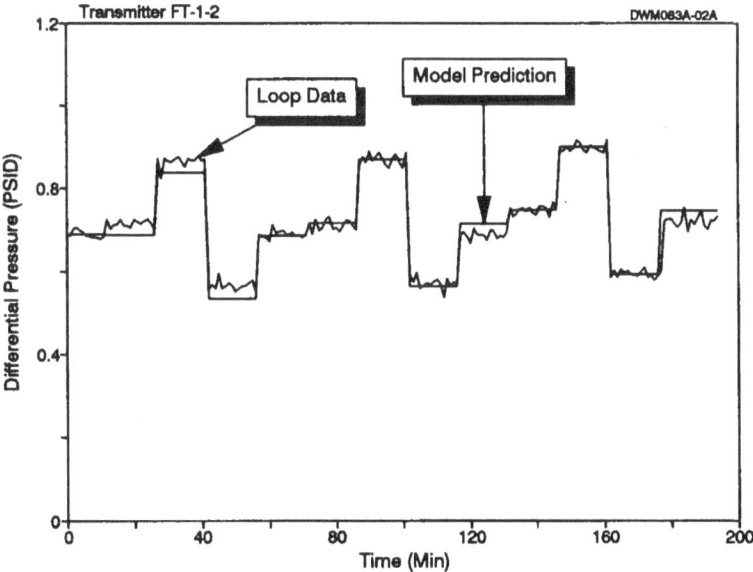

Figure 10.2   Comparison between empirical modeling results and measured data for two flow transmitters

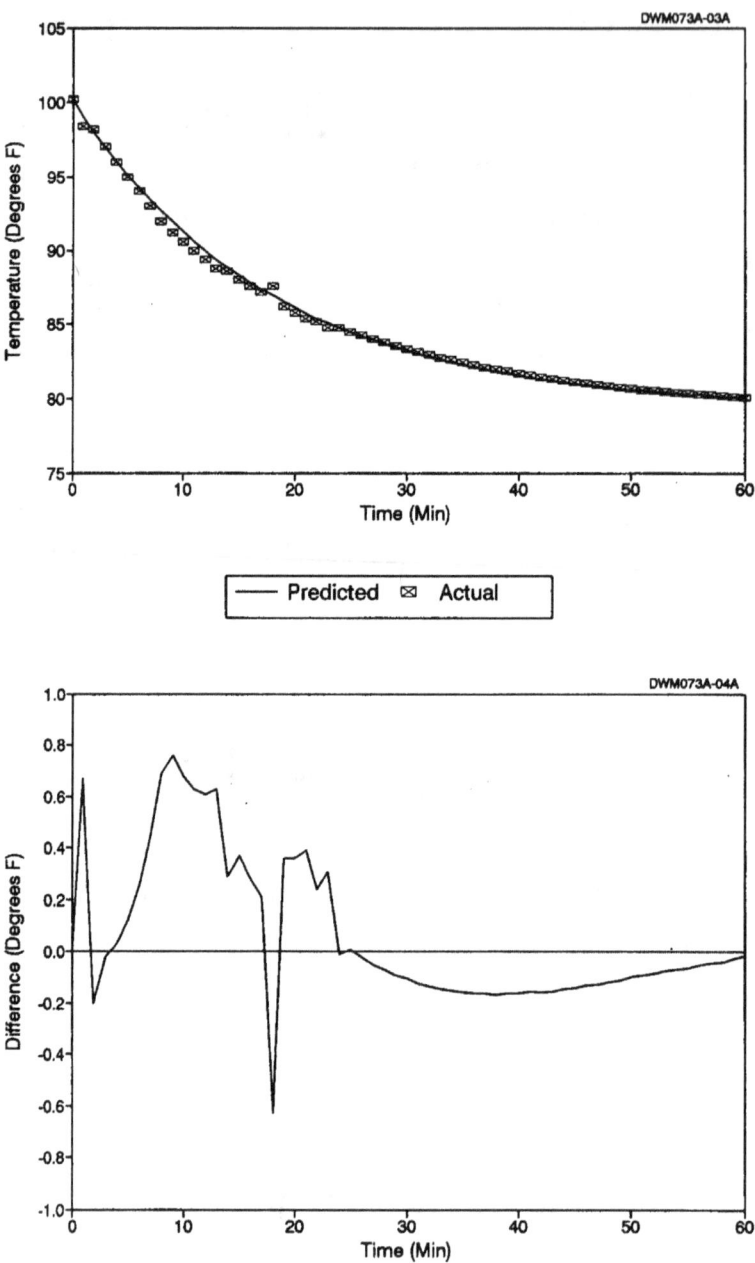

Figure 10.3   Agreement between results of physical model and
actual temperature data

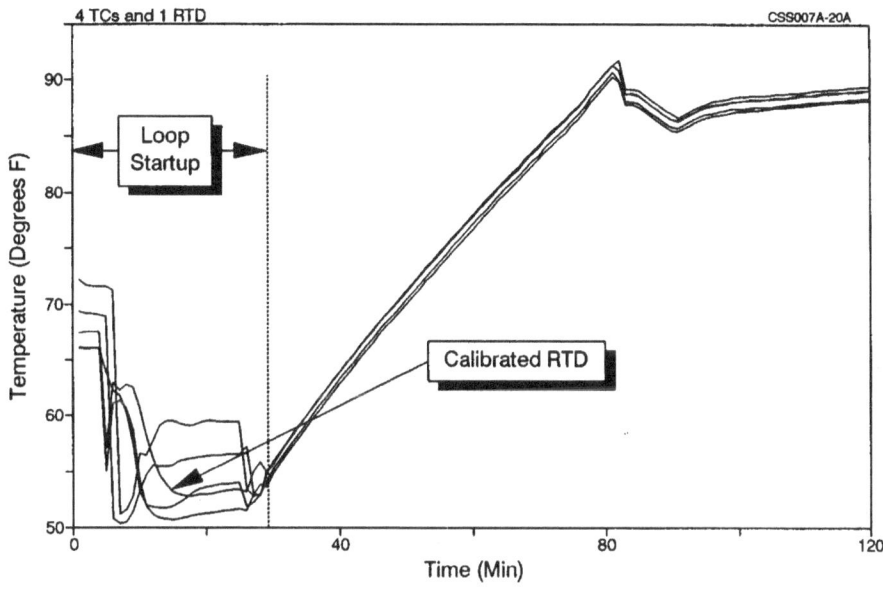

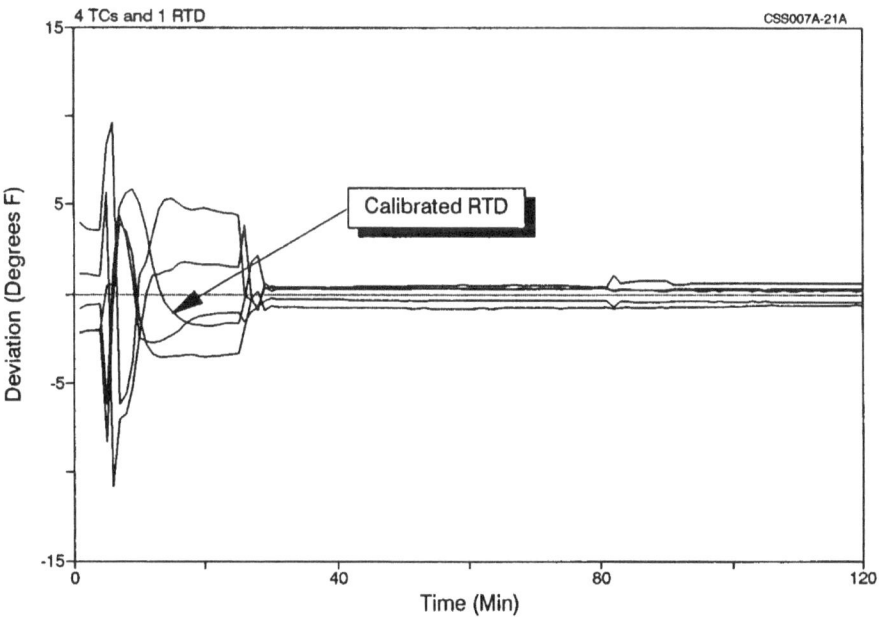

Figure 10.4 Agreement between five temperature sensors monitoring the
loop temperature from start up to steady-state

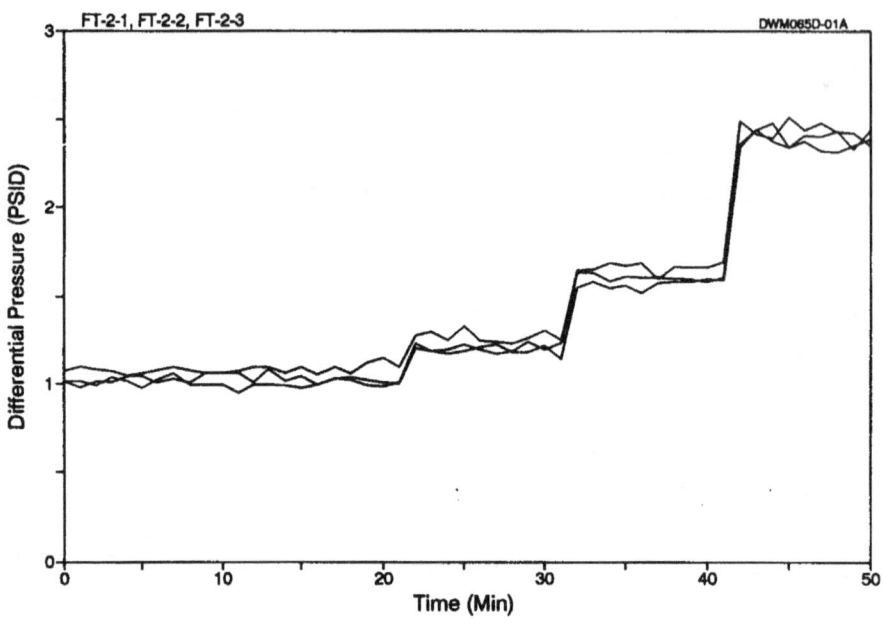

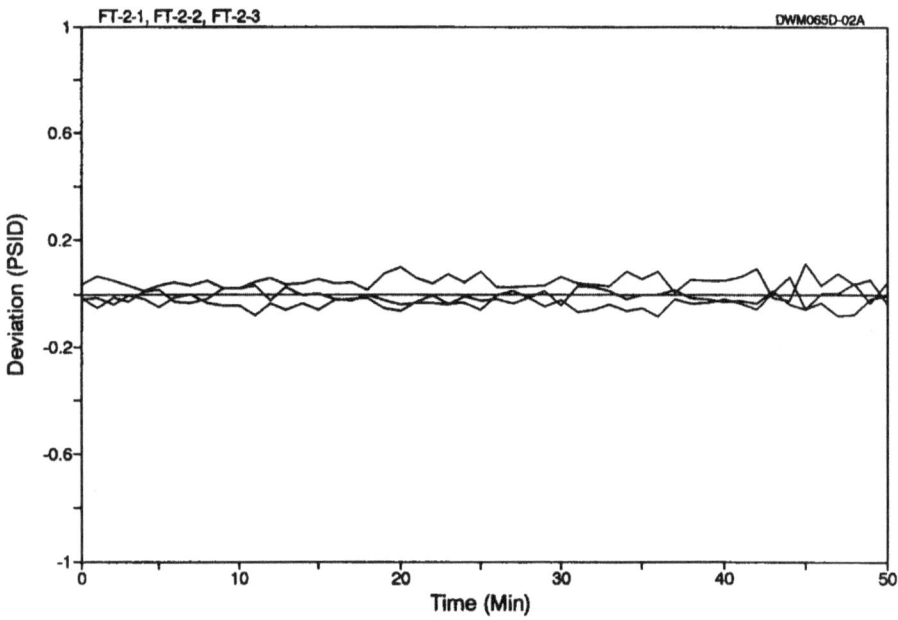

Figure 10.5 Monitoring data for three flow transmitters

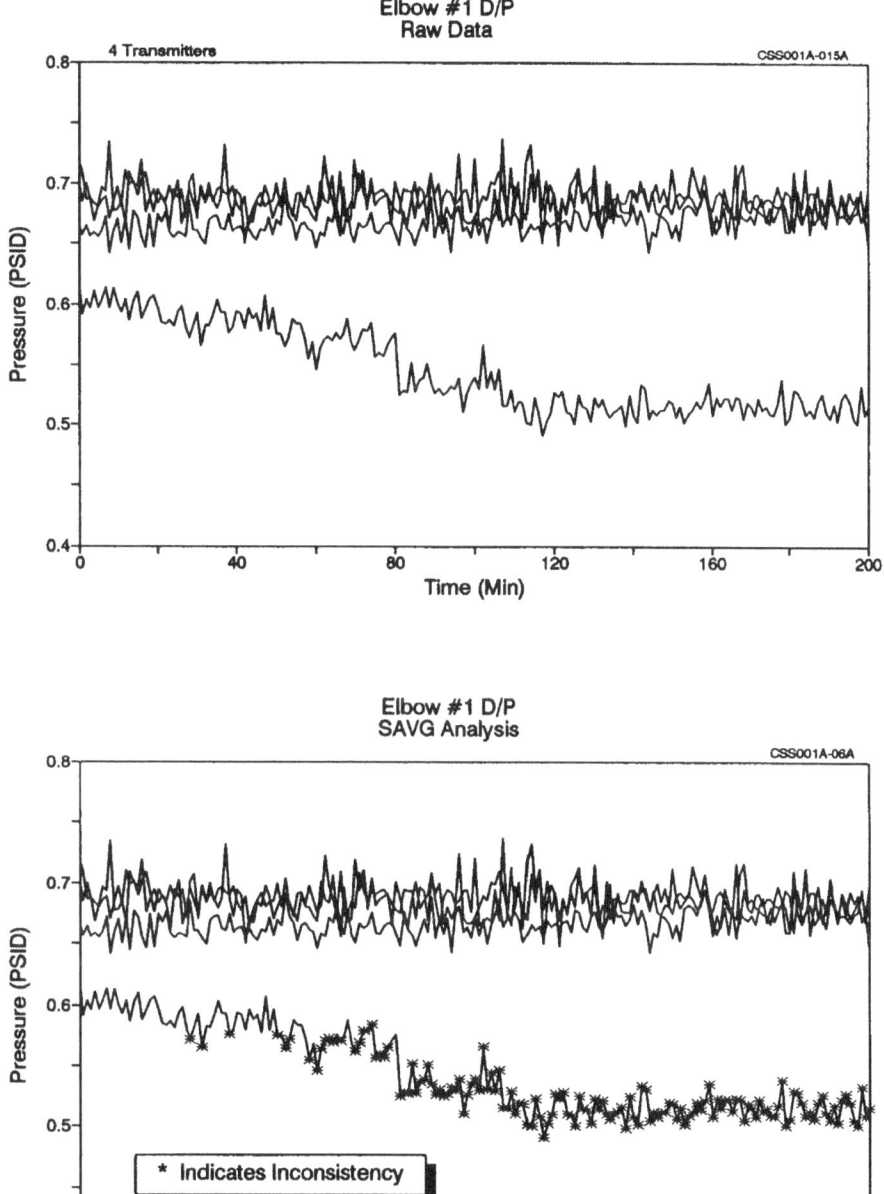

Figure 10.6   Demonstration of consistency checking with four differential pressure signals

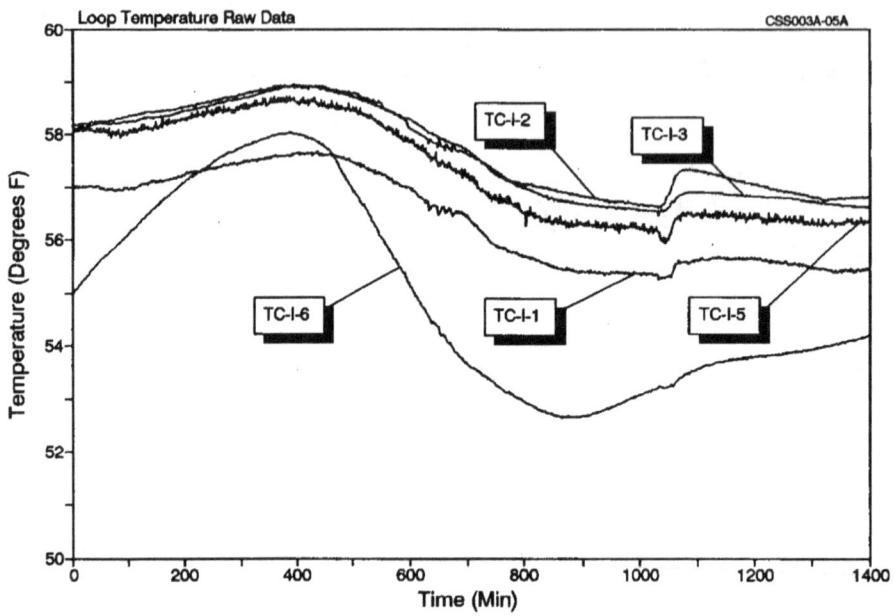

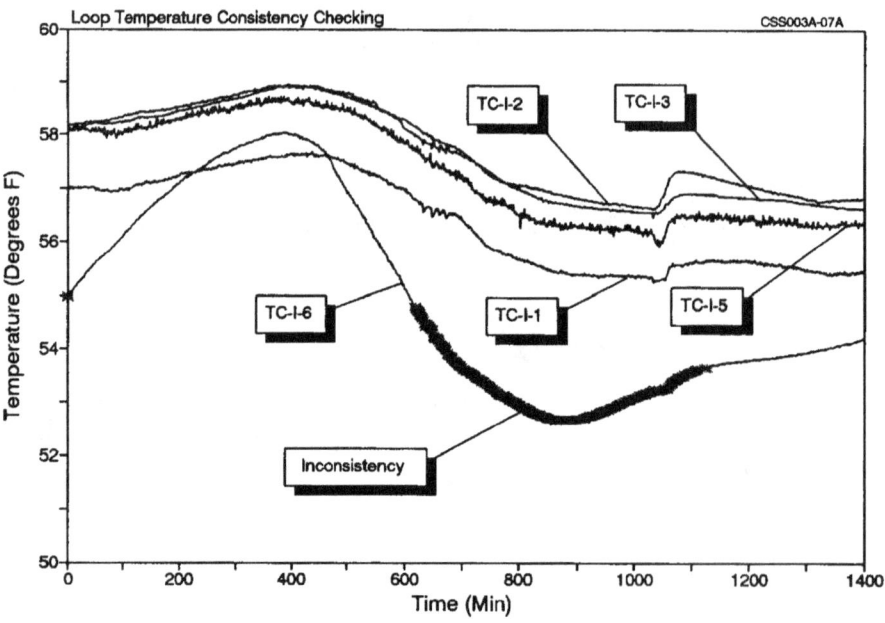

Figure 10.7 Consistency checking of temperature signals

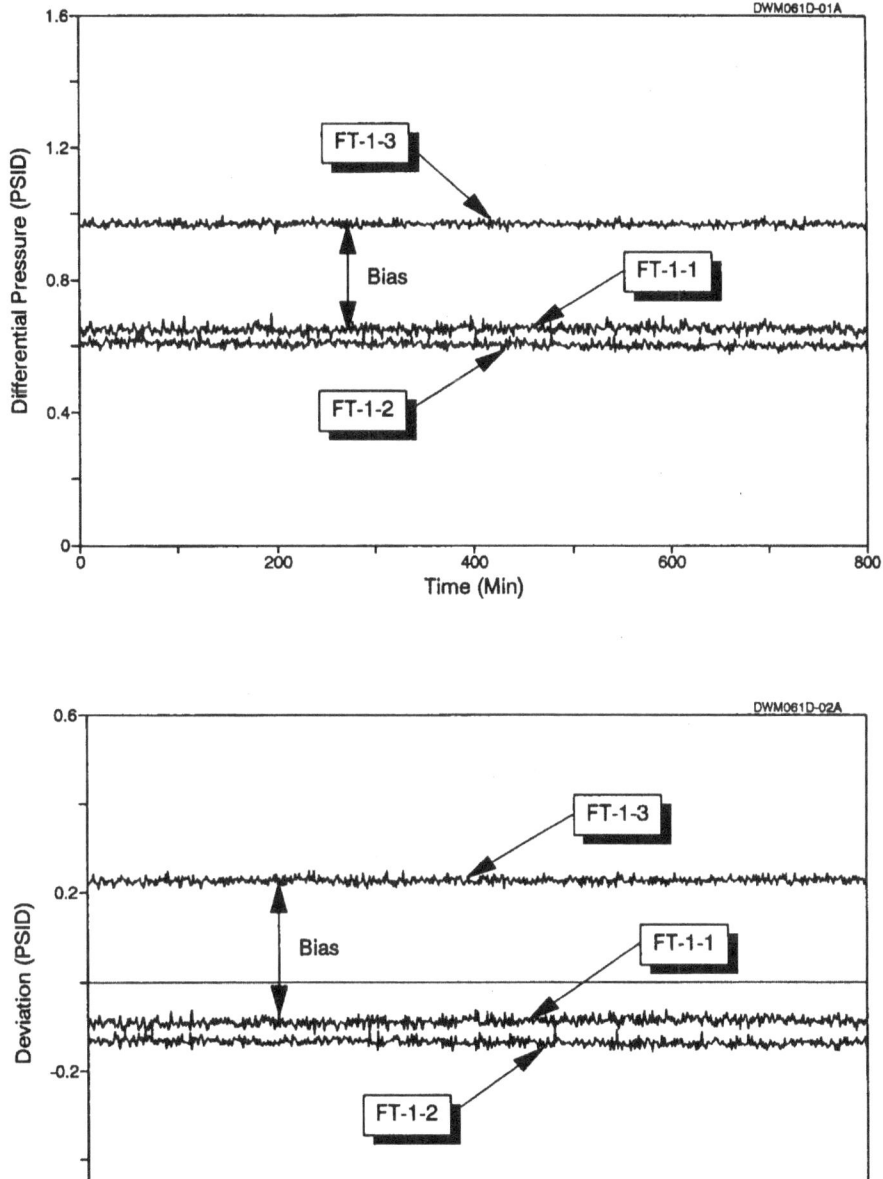

Figure 10.8 Flow transmitter data with induced bias

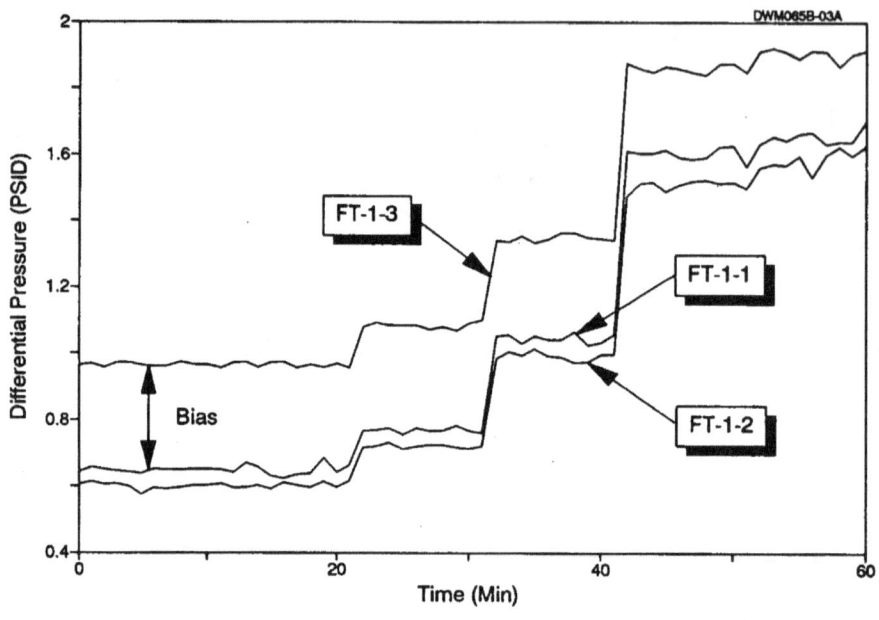

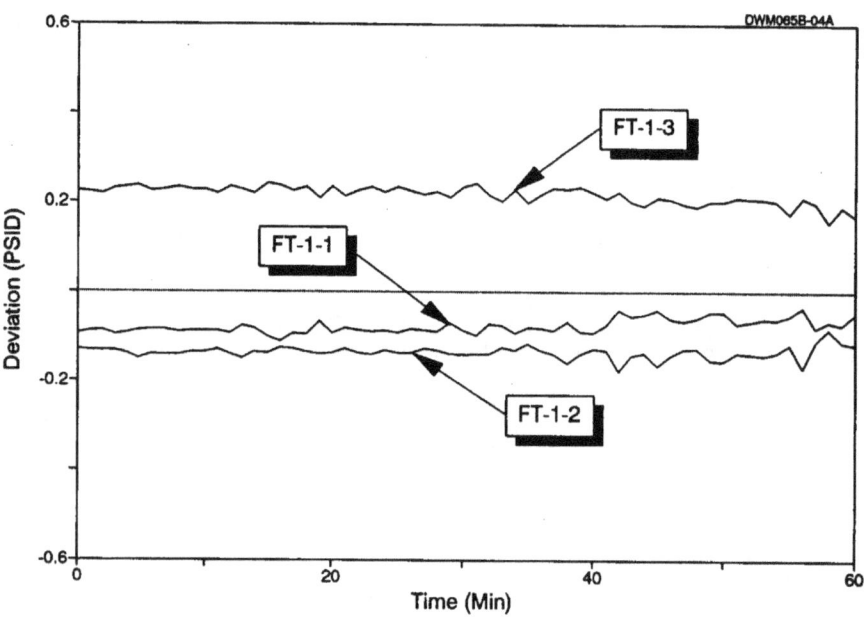

Figure 10.9   Detection of bias by on-line monitoring

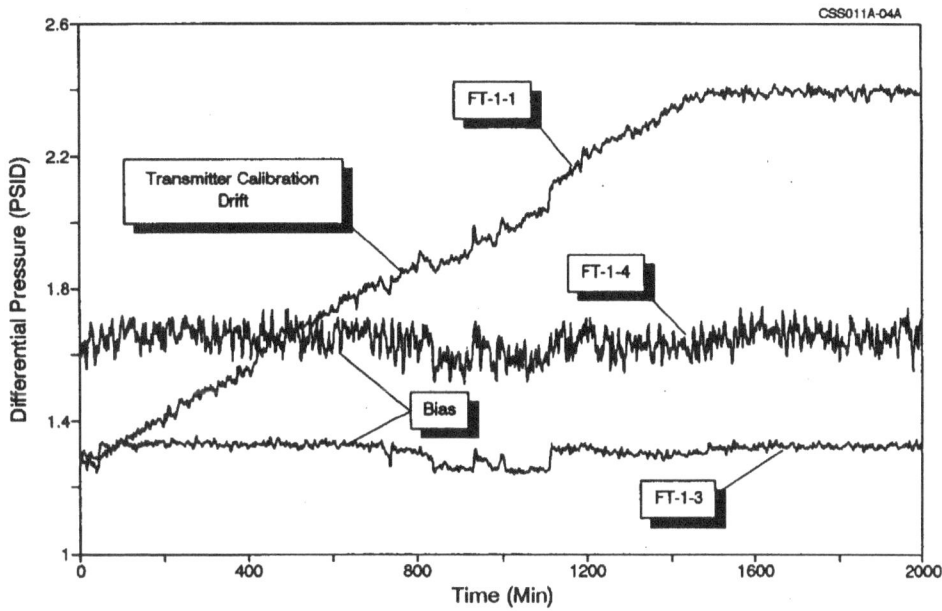

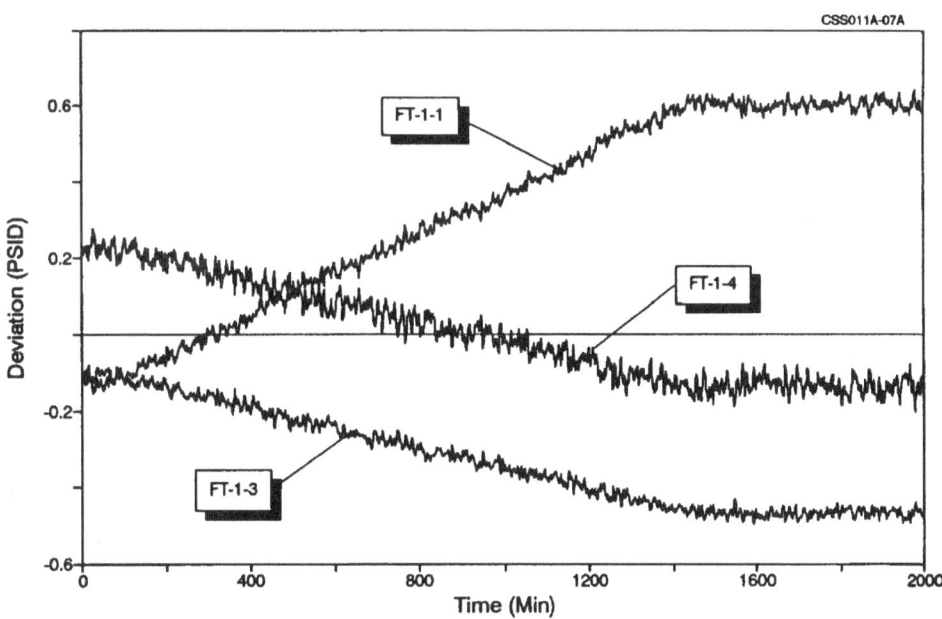

Figure 10.10   Bias and drift in transmitter signals

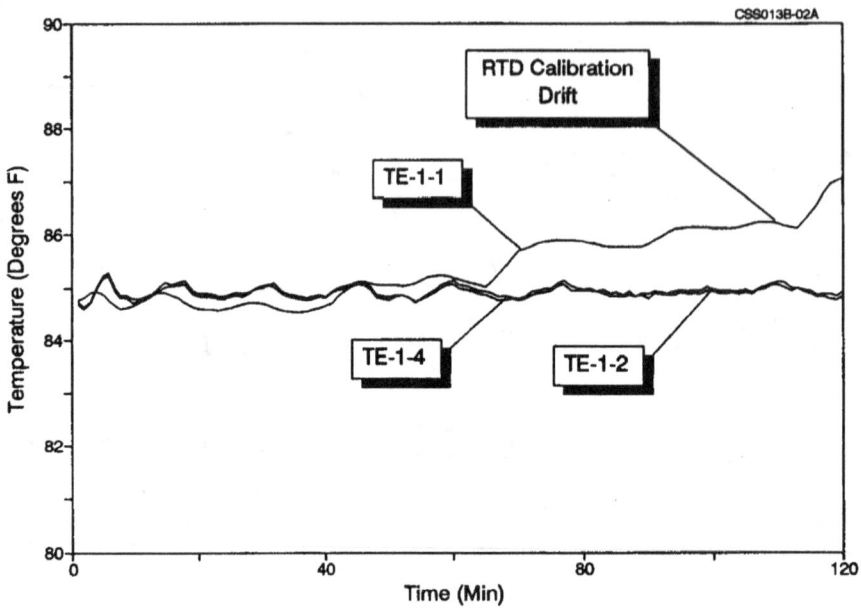

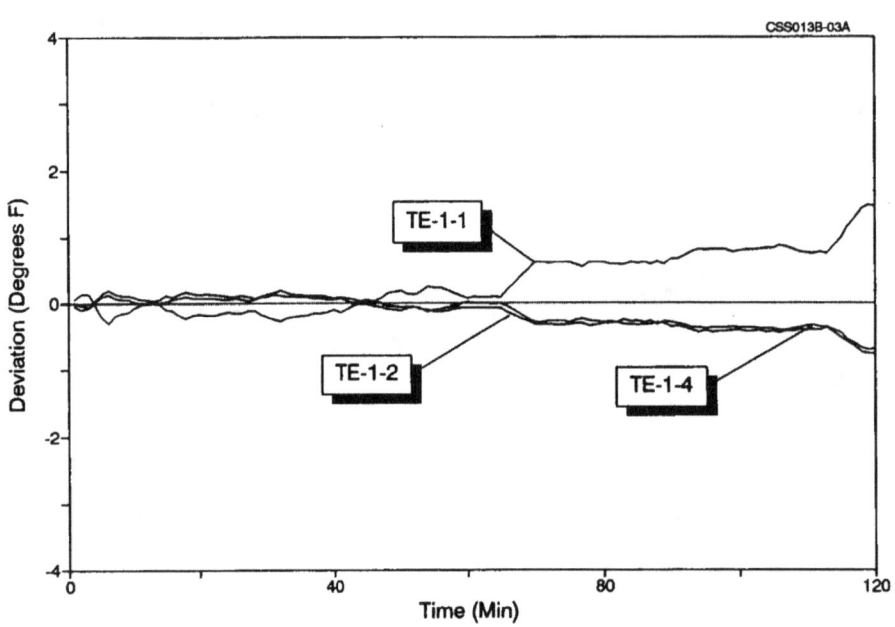

Figure 10.11 Bias and drift induced in temperature signals

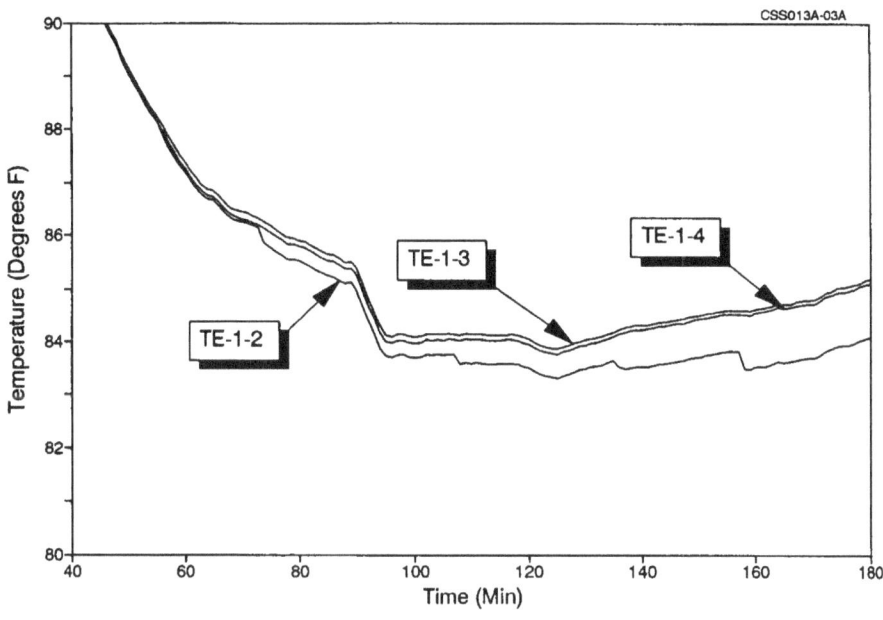

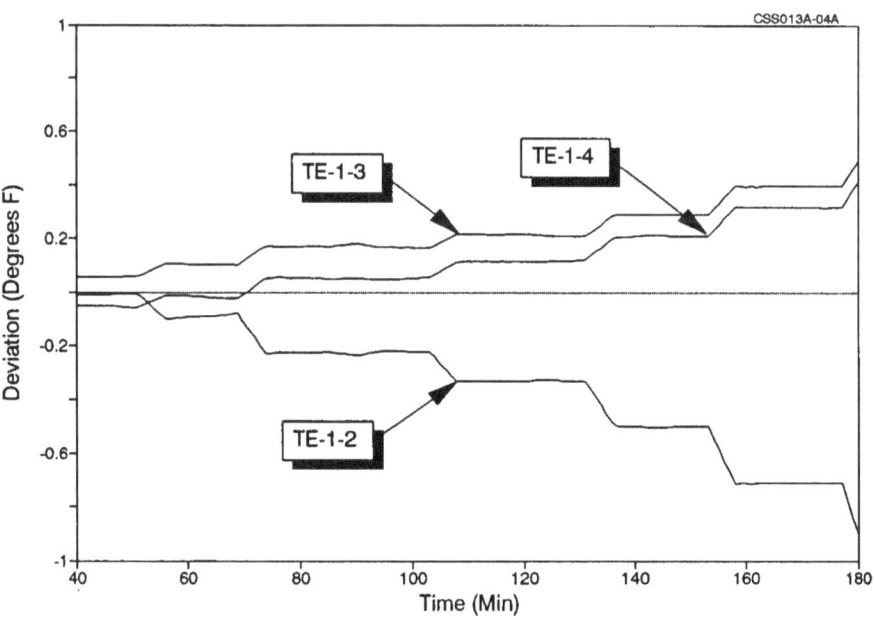

Figure 10.12   Temperature signals with induced drift

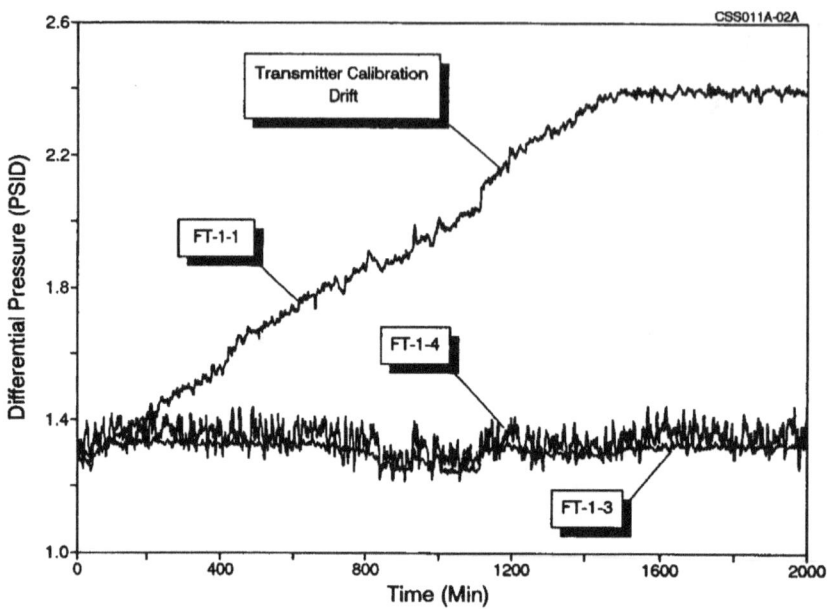

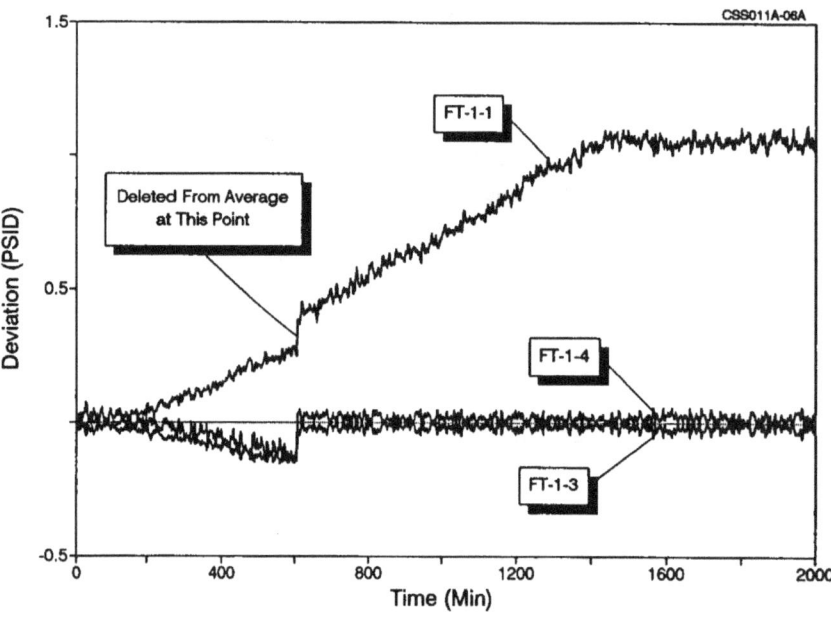

Figure 10.13   Demonstration of a sudden shift in deviation plot when a
signal becomes excluded from the average

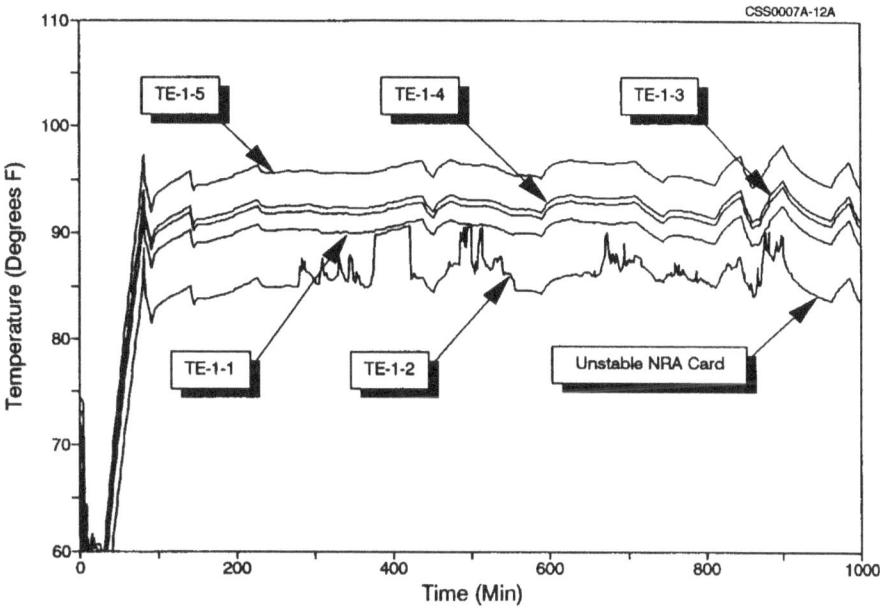

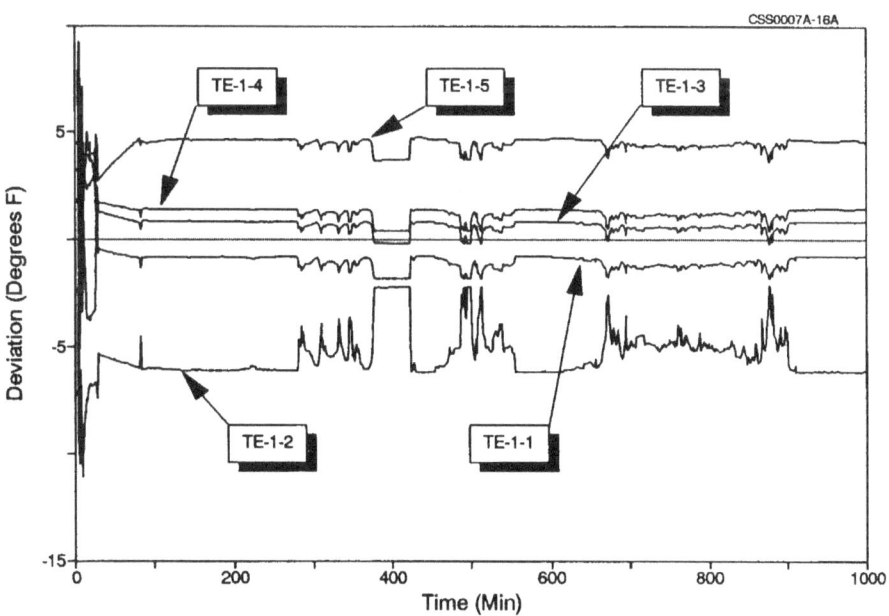

Figure 10.14   Effects of an erratic signal on deviation plots of five
temperature sensors

We should point out once again that the purpose of generating the data and the plots discussed above and many others that were generated in the course of laboratory tests reported herein has been to provide a database of information to serve as a learning tool in developing the experience that is necessary to interpret the behavior of various redundant and non-redundant signals under a variety of conditions.

In addition to studies with DC data, laboratory tests were performed using AC data (noise) to demonstrate how response time degradation is identified by on-line monitoring. The AC data available from the test loop are the result of flow fluctuations due to turbulence in the loop. The AC data were processed with various techniques including FFT, AR, and zero crossing analyses. Figure 10.15 shows a block diagram of the programs that were used to analyze the AC data. An example of representative results of analyses of laboratory noise data is shown in Figure 10.16 in terms of PSDs obtained from FFT and AR analyses. The data in this plot represent the dynamic response characteristics of a Barton transmitter installed in the test loop. The transmitter was first tested in a normal configuration and then tested with an induced degradation to cause it to have a slower dynamic response. The degradation was induced by introducing a partial blockage in the sensing line leading to the transmitter. The degradation manifests itself in the PSD results by shifting the roll-off frequency of the PSD to a lower frequency region. A similar plot is shown in Figure 10.17 for a Rosemount pressure transmitter. This transmitter is equipped with a damping adjustment

that is normally used to reduce the high frequency noise at the output of the transmitter. The damping adjustment was manipulated to slow the transmitter's response and demonstrate that the problem can be identified in the results of the noise analysis. Additional plots of AC tests are given in Figures 10.18 through 10.20 for transmitters from three other manufacturers of nuclear grade pressure transmitters.

The two PSD plots on the bottom of Figures 10.16 to 10.20, which are from the AR analysis, appear cleaner than the two FFT PSDs on the top. This is because FFT analysis produces the PSD directly from the data while the AR results are generated from the AR model after the noise data are fit to the AR model and the model parameters are identified and used to plot the PSD. Furthermore, in producing the AR results plotted here, the AR model was forced to use a model order of 5, causing the PSD to appear smoother by excluding the small resonances and the harmonics from the plot.

For a Foxboro pressure transmitter tested in the laboratory, instead of FFT or AR analysis, the noise data were used to obtain a zero crossing rate for the transmitter. The zero crossing rate was identified for a normal and an artificially degraded configuration. The results are shown in Figure 10.21 in terms of an initial and a final zero crossing rate for the normal and degraded configurations, respectively.

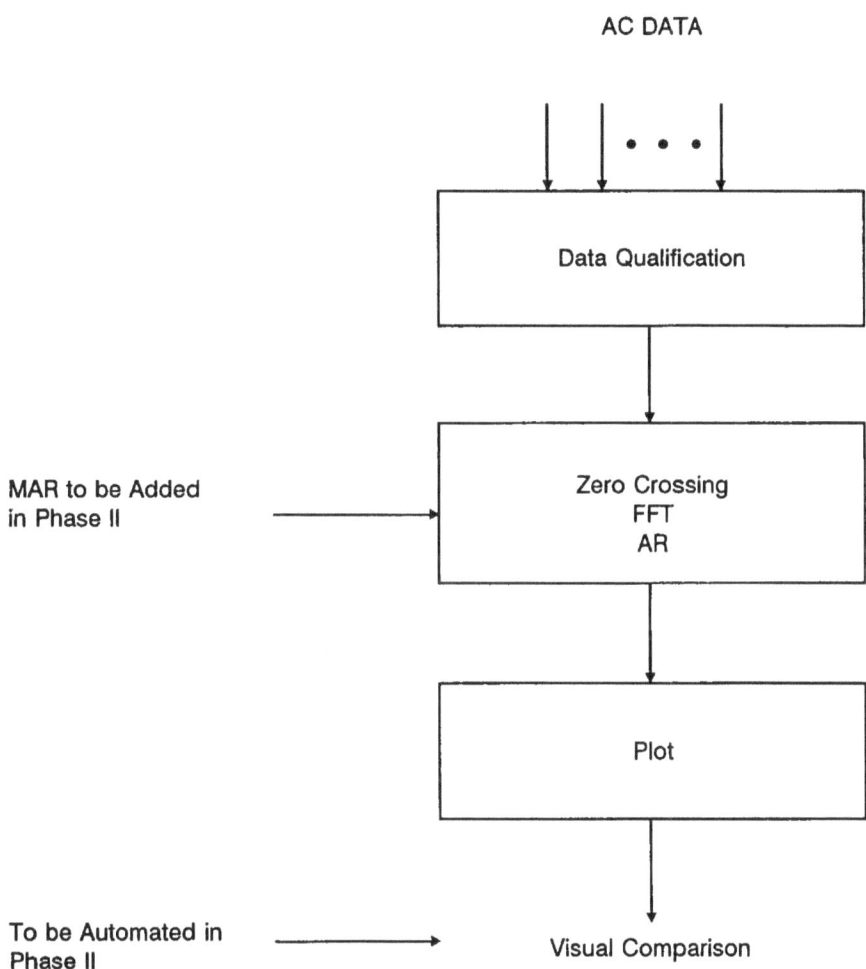

Figure 10.15   Steps in AC data analysis

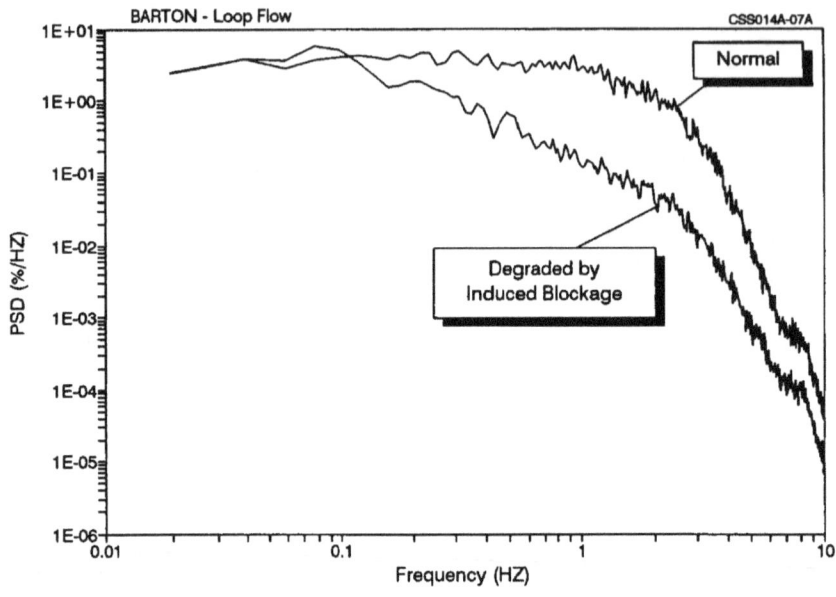

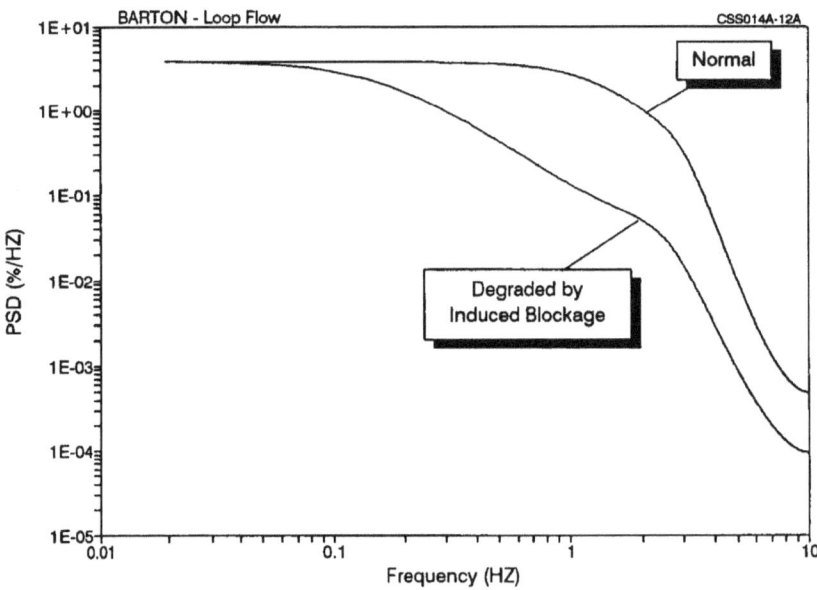

Figure 10.16   PSD plots from FFT (top) and AR (bottom) for a Barton transmitter before and after an induced degradation in dynamic response

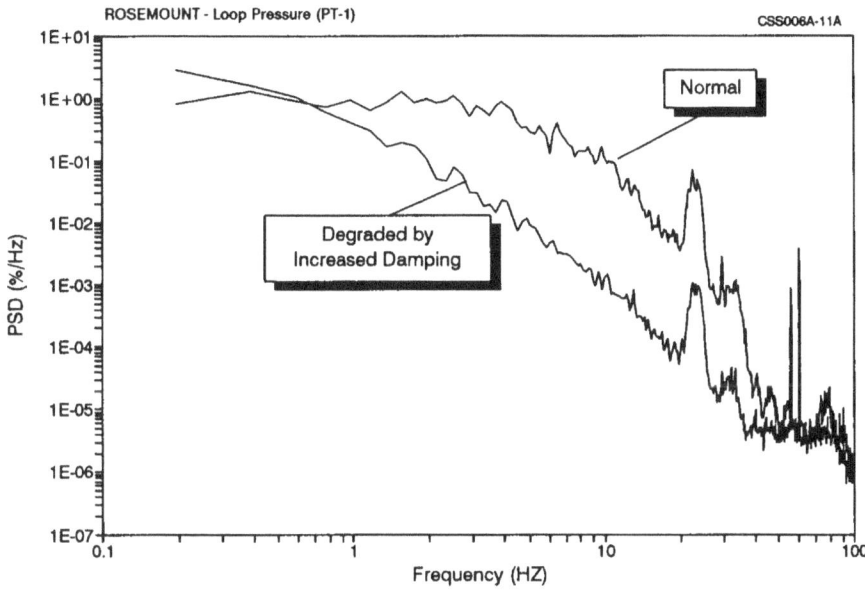

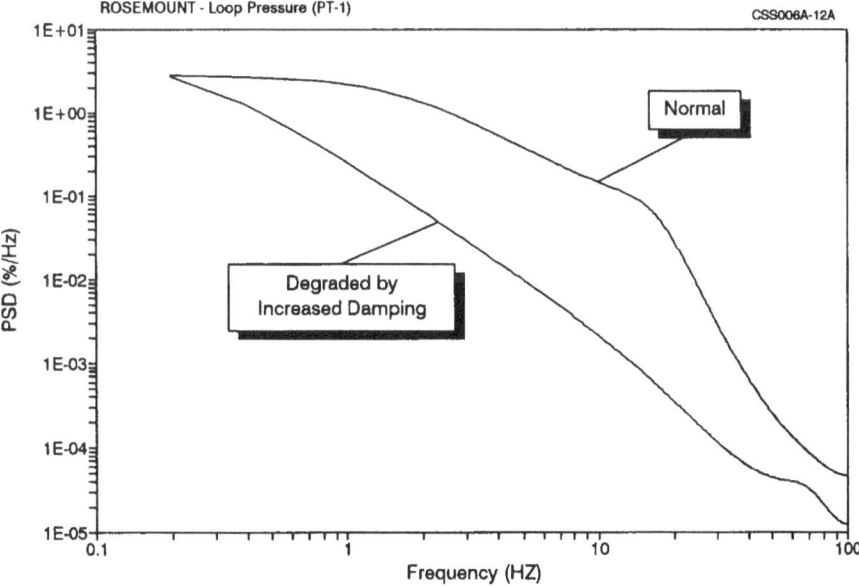

Figure 10.17   PSD plots from FFT (top) and AR (bottom) for a Rosemount transmitter before and after induced degradation in dynamic response

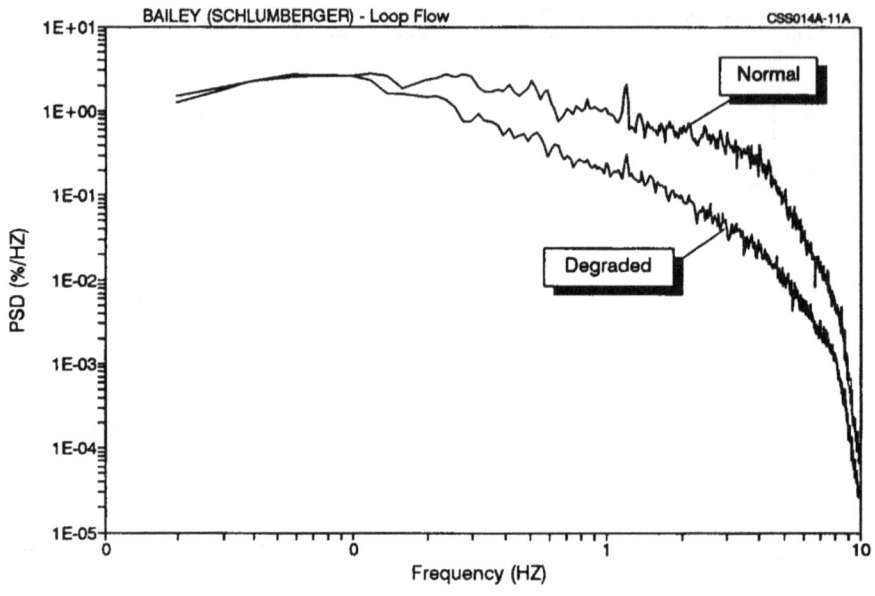

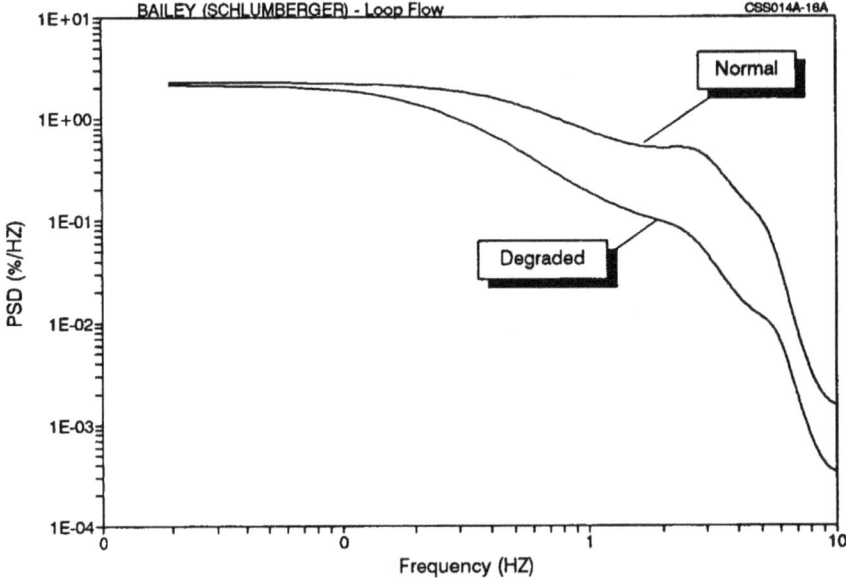

Figure 10.18   PSD plots from FFT (top) and AR (bottom) for a Schlumberger transmitter before and after induced degradation in dynamic response

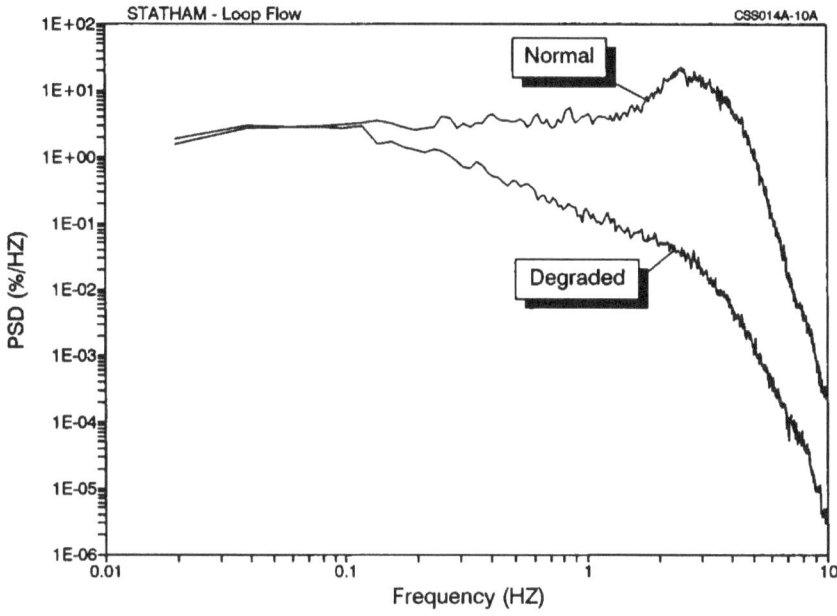

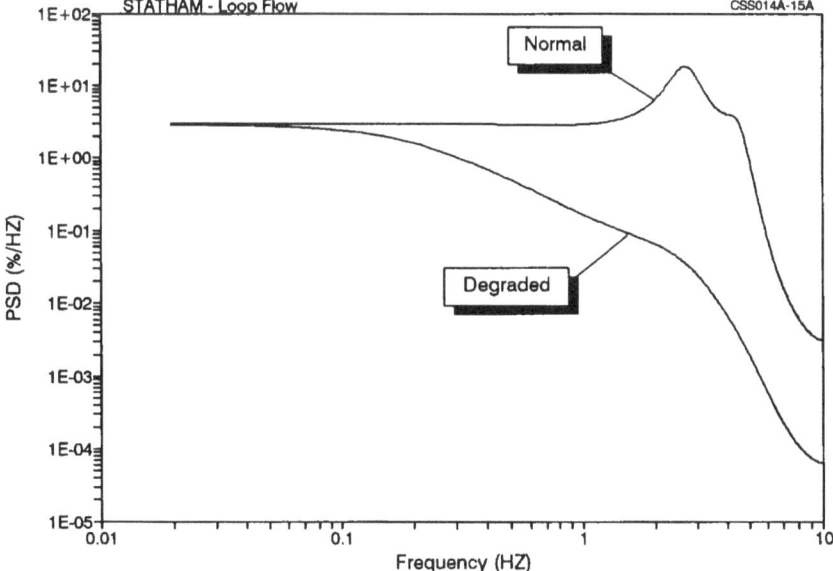

Figure 10.19 PSD plots from FFT (top) and AR (bottom) for a Statham transmitter before and after induced degradation in dynamic response

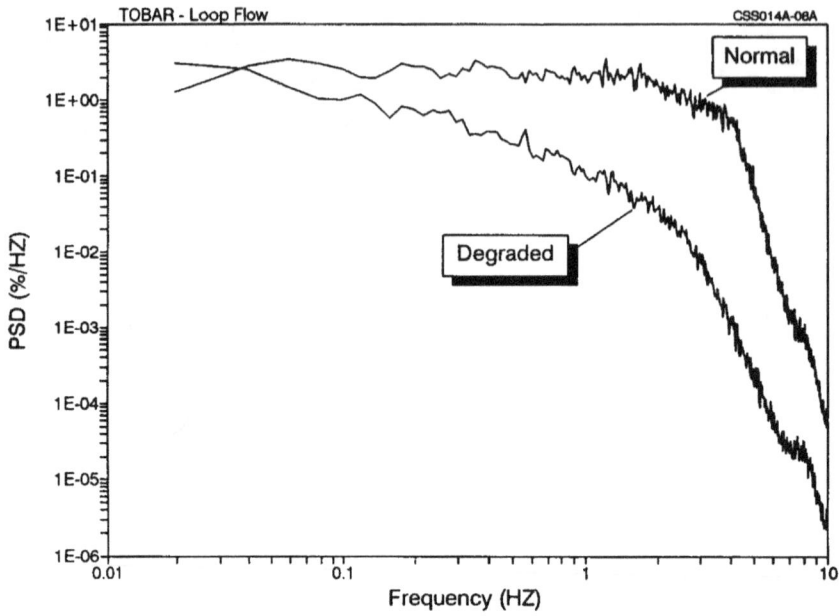

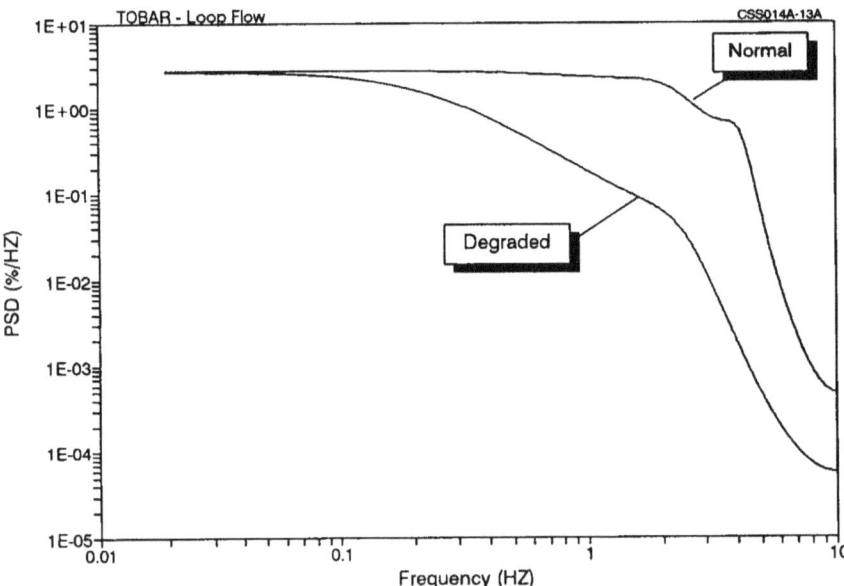

Figure 10.20 PSD plots from FFT (top) and AR (bottom) for a Tobar transmitter before and after induced degradation in dynamic response

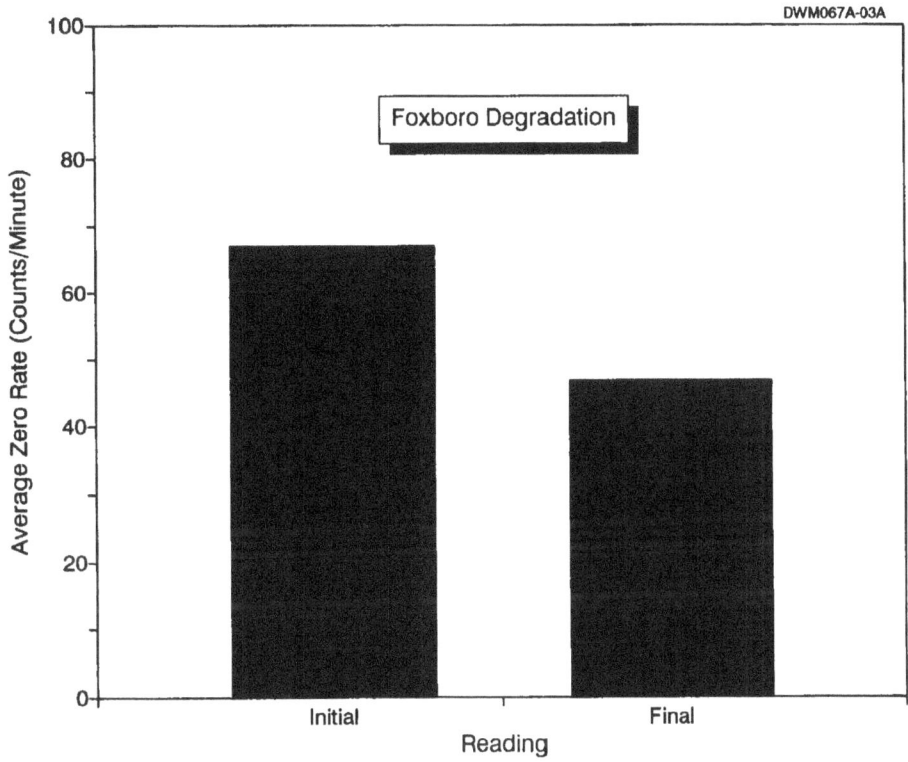

Figure 10.21 Zero crossing rate for a Foxboro transmitter before and after an induced degradation in response time

11. In-Plant Test Results

The in-plant validation of smart sensor technologies is the most important aspect of this project. The in-plant validation work is being conducted at the McGuire Nuclear Station Unit 2 where 170 signals from the primary and secondary systems of the plant are being monitored continuously, including when the plant is at cold shutdown. The monitoring began in March 1992 at the beginning of the plant's eighth fuel cycle and will continue for two complete fuel cycles. A listing of the types of signals being monitored is given in Table 11.1. This is followed by a drawing of one of the four loops of the McGuire plant in Figure 11.1. The figure shows the sensors that are monitored in each of the four loops. These include both safety-related and non-safety-related sensors. An attempt has been made to use all the redundant sensors from each service, but in some cases only two out of a group of three or four sensors were available for monitoring. It should be pointed out here before proceeding to present in-plant results that the work at the McGuire plant and the test results given here are of a preliminary research nature and should not be used as a basis for any conclusions without further investigation.

Figure 11.2 shows a block diagram of a typical instrument loop for a flow transmitter and how the flow signal is connected to the on-line monitoring system. It is clear that the monitoring includes not only the sensor (pressure, level, and flow transmitter, RTD, thermocouple, neutron detector, etc.), but also the instrument loop electronics. However the logic and trip circuitry are not necessarily included in the monitoring. A more detailed drawing of an instrument loop is shown in Figure 11.3 indicating the locations of the isolated outputs (ISO) where the on-line monitoring system may be connected.

The main purpose of the in-plant validation effort is to determine if the on-line monitoring system is able to identify the same channels as faulty as those identified as faulty by the regular surveillance tests and the off-line calibrations. A fault could be due to an erratic behavior such as large spikes, noise, a sudden shift, or a calibration drift or response time increase. An inherent problem (from a research point of view) with any in-plant validation of this type is that there are very few instrument channels in a nuclear power plant that show significant drift or degradation during a fuel cycle. More specifically, there are very few faults to be detected by the on-line monitoring system and used for the validation tests. Typically, of a hundred channels that

are tested periodically in a nuclear power plant, less than ten are found to be out of tolerance or have any fault. As a matter of good practice, however, plant technicians often null any deviation observed during the surveillance and calibration tests even if the deviation is well within the acceptance band. This point is manifested in a comparison between the on-line monitoring results and the surveillance test records at McGuire (see Figures 11.4 through 11.6). In each of the three figures, two plots are shown as follows: 1) deviations of each signal from the average of the redundant signals in the group, and 2) the plot of the signal that was adjusted during the monthly surveillance tests. Note that in all three cases, the channel that was adjusted is the same channel that was identified by the on-line monitoring system as having the largest deviation.

Comparisons of the types discussed above will be performed throughout the in-plant validation tests in Phase II, especially with data from full channel calibrations performed during refueling outages. More data on the McGuire pressure transmitters is included later in this chapter. However, before proceeding to discuss other sensors, it is important to show a possible behavior in the results of analysis of on-line monitoring data that is not related to the sensor or other components of a channel. Figure 11.7 shows three deviation traces for redundant flow signals at McGuire. Until about 120 days into the data, the NCFT5040 is excluded from the average because of its large deviation from the other two redundant transmitters. That is, only two of the three signals are averaged to obtain the best estimate of the process parameter. As shown in Figure 11.7, this causes the two traces around the zero line to appear as mirror images of one another. After about 120 days, NCFT5040 drifts upward and becomes close to being consistent enough to be included in the average. At this point, the signals are compared against the average of the three traces and the mirror images start to disappear. Note the repeated spikes in the three traces after 120 days. These spikes are not in the signals. Rather, they are due to the fact that NCFT5040 continues to bounce in and out of the average and shift the level of the three signals with respect to the average. Another example of such an effect is shown in Figure 11.8 for four redundant steam generator level signals.

The next interesting point about the in-plant tests is related to the core exit thermocouples (CETs). Of the

Table 11.1

Listing of Signals Monitored at McGuire Unit 2

Item	Description of Signal	Number of Signals
1	Steam Flow	8
2	Steam Pressure	12
3	Steam Generator Level	20
4	Feedwater Flow	8
5	Auxiliary Feedwater Flow	4
6	Reactor Coolant Flow	12
7	Pressurizer Level	3
8	Pressurizer Pressure	4
9	Wide Range Reactor Coolant Pressure	2
10	Containment Pressure	3
11	Reactor Vessel Level Indicating System (RVLIS)	6
12	Turbine Impulse Pressure	2
13	Neutron Flux Detectors (NI Channels)	12
14	Narrow Range RTDs	16
15	Wide Range RTDs	8
16	Core Exit Thermocouples	40
17	Miscellaneous	10
	Total Signals	**170**

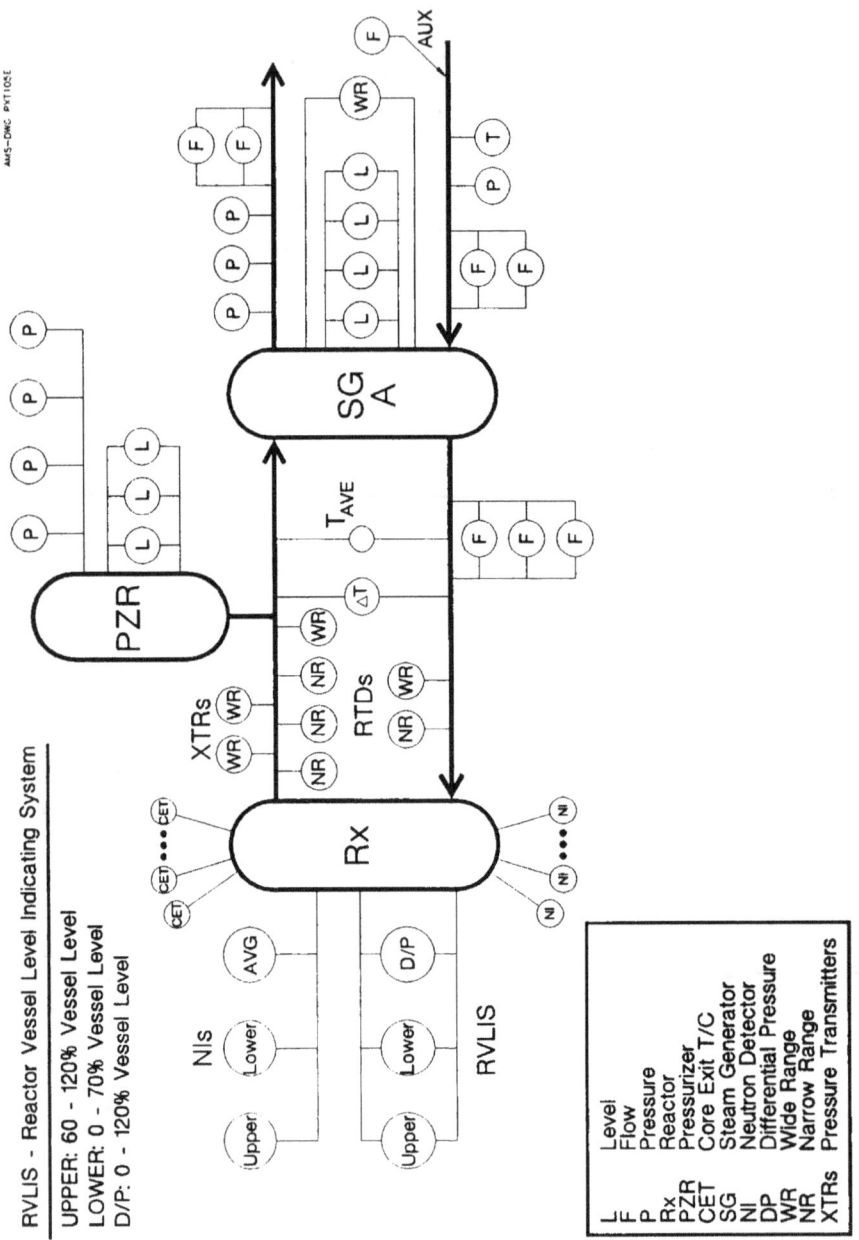

RVLIS - Reactor Vessel Level Indicating System

UPPER: 60 - 120% Vessel Level
LOWER: 0 - 70% Vessel Level
D/P: 0 - 120% Vessel Level

L	Level
F	Flow
P	Pressure
Rx	Reactor
PZR	Pressurizer
CET	Core Exit T/C
SG	Steam Generator
NI	Neutron Detector
DP	Differential Pressure
WR	Wide Range
NR	Narrow Range
XTRs	Pressure Transmitters

Figure 11.1   Sensors being monitored in one of the four loops of the McGuire plant

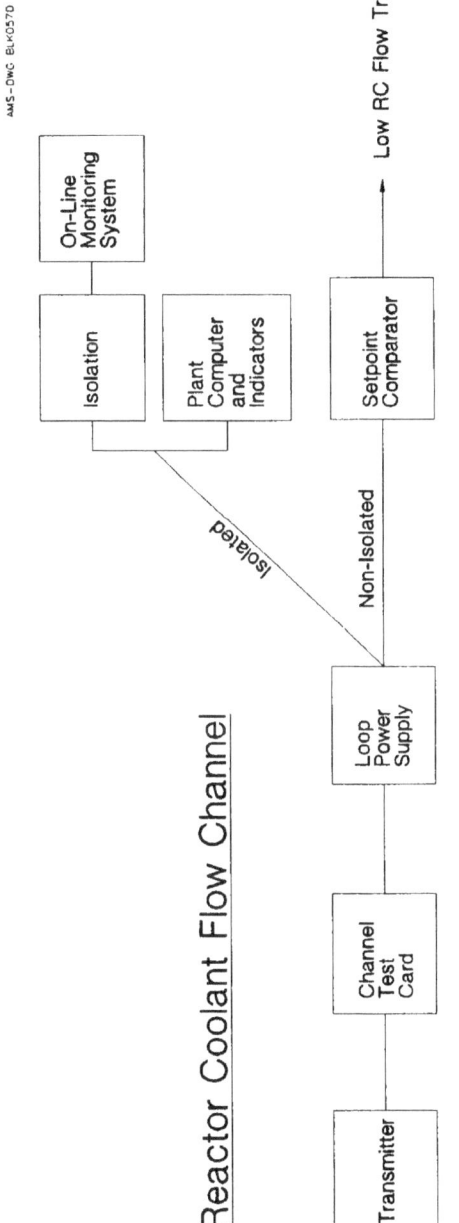

Figure 11.2 Placement of the on-line monitoring system in an instrument channel

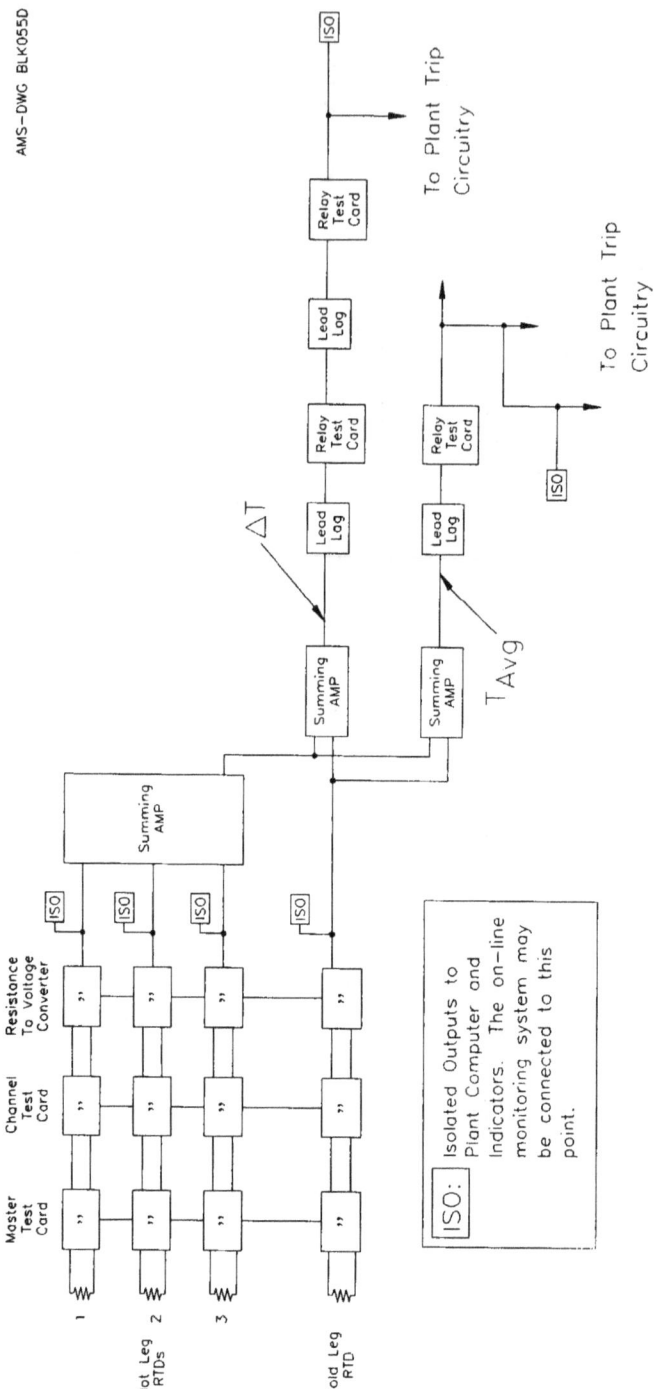

AMS–DWG BLK055D

ISO: Isolated Outputs to Plant Computer and Indicators. The on-line monitoring system may be connected to this point.

Figure 11.3 **Simplified schematic of a typical redundant instrument set and isolated outputs where signals are typically available for on-line monitoring**

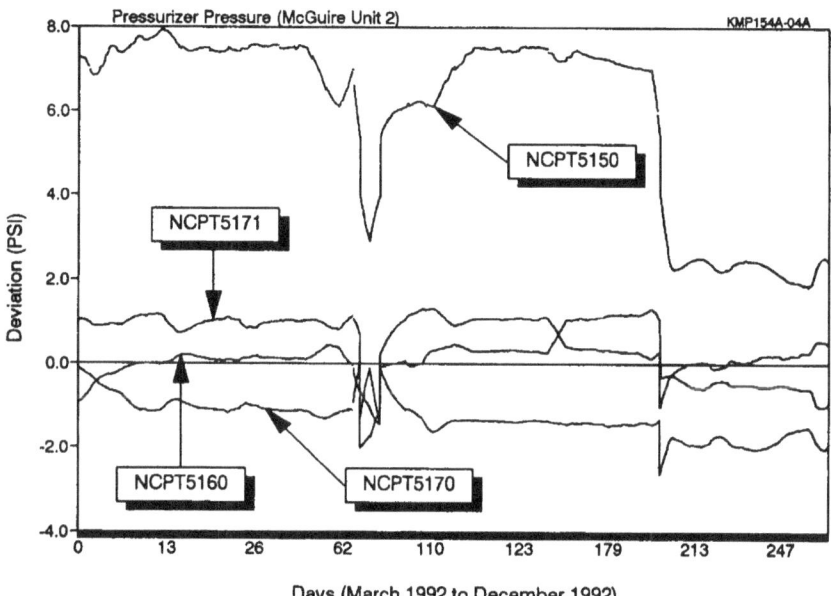

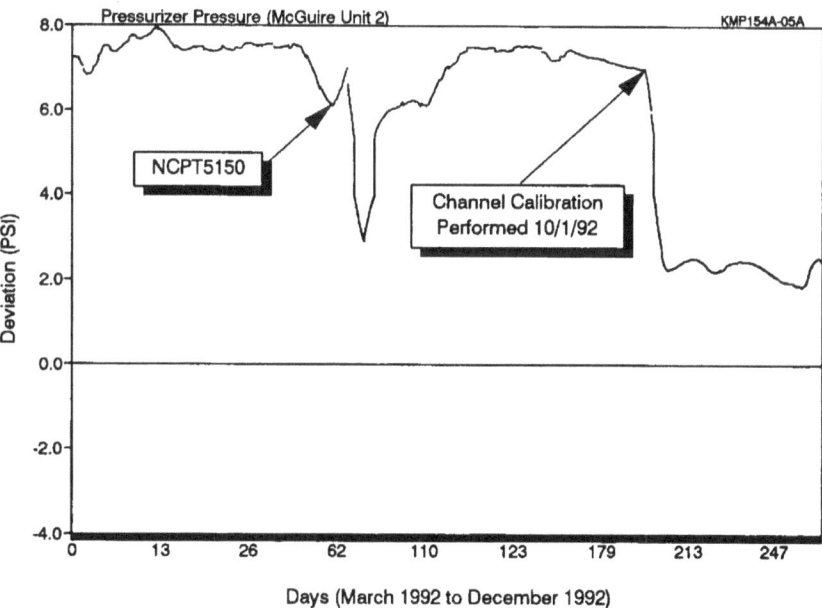

Figure 11.4 Detection of a deviation in a pressurizer pressure signal that
was nulled during a surveillance test

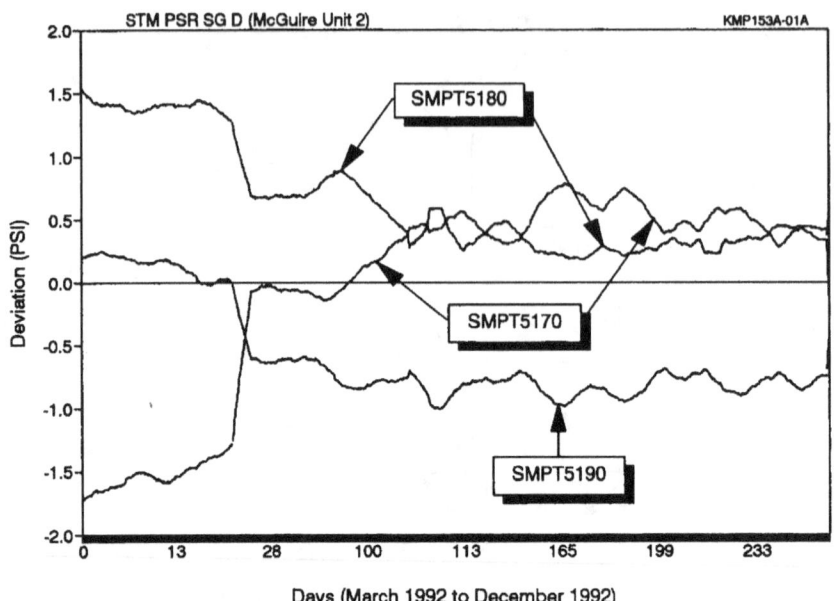

Days (March 1992 to December 1992)

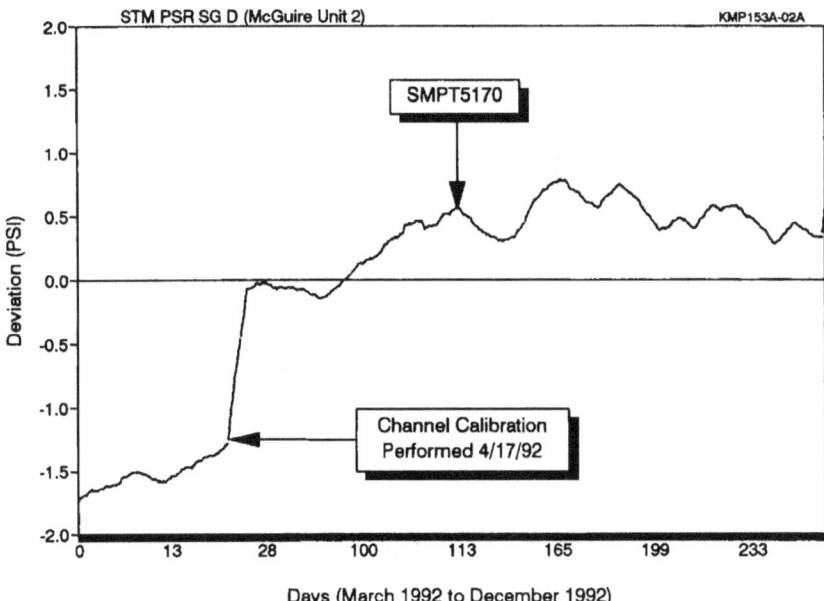

Days (March 1992 to December 1992)

Figure 11.5   Detection of a deviation in a steam pressure signal that was
nulled during a surveillance test

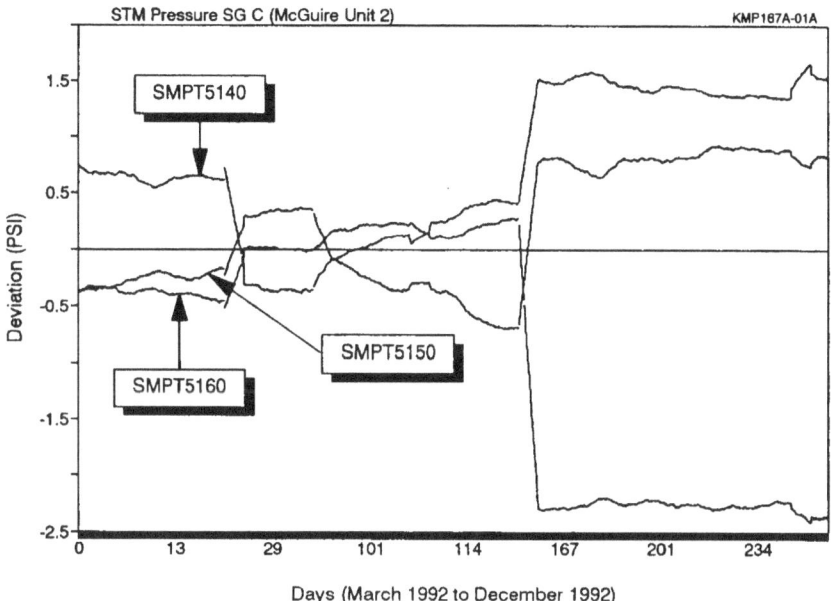

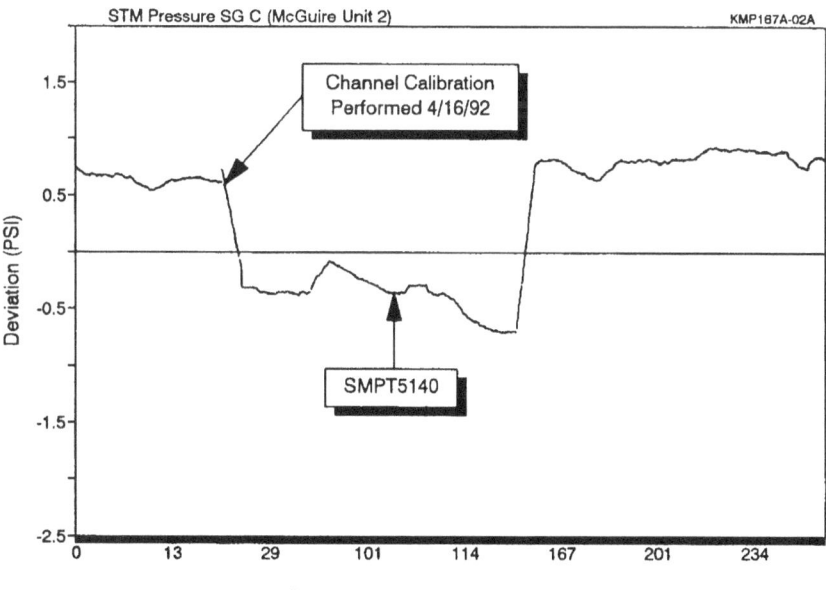

Figure 11.6 Detection of a deviation in a steam pressure signal that was nulled during a surveillance test

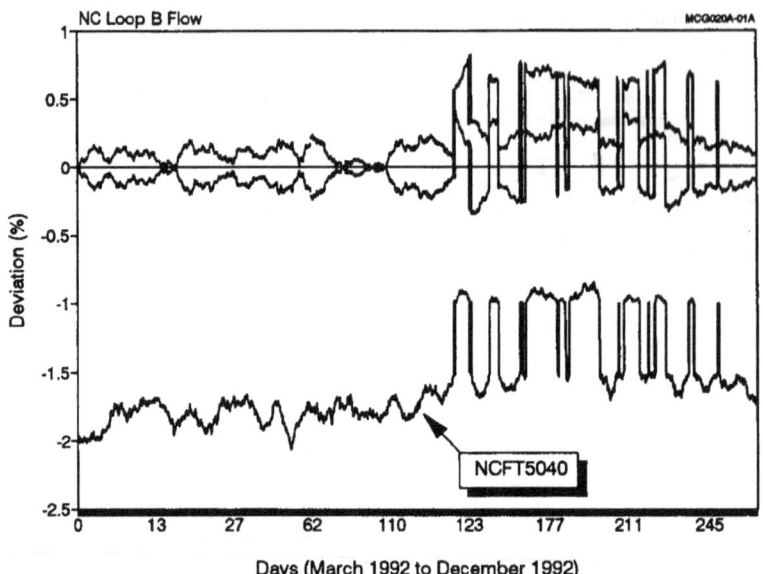

Figure 11.7   Demonstration of averaging spikes in three flow signals

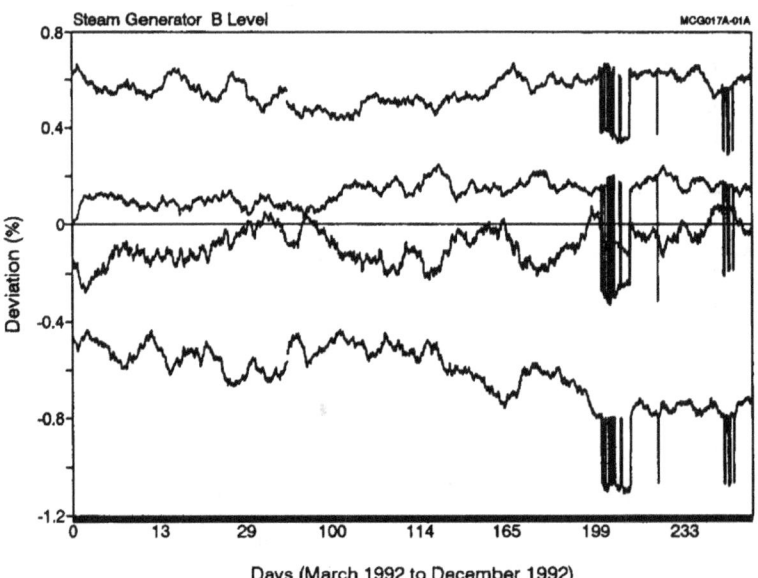

Figure 11.8   Demonstration of averaging spikes in four level signals

- 130 -

forty thermocouples being monitored at McGuire, thirty-seven have useful signals. These signals can be considered as redundant because they should be measuring the same temperature with the exception of any temperature streaming due to the water not being well mixed as it exits the core. Table 11.2 shows the results of cross calibration tests of the CETs at full power and shutdown conditions. It is apparent that the thermocouples have much smaller errors at ambient temperature ($\sim$110°F) than at operating temperature ($\sim$620°F). The data in Table 11.2 represent the deviations of the individual thermocouples from the average of all 37 thermocouples. For the results at power, the outliers (those with deviations greater than $\pm$25°F) were removed in determining the average temperature, but for the shutdown results, the outliers were not removed because their deviations were generally small. The deviation criteria of 25°F is based on the expected accuracy of thermocouples, temperature streaming effects, and measurement uncertainties. The temperature data used for the cross calibration of CETs were the average of approximately 60 points sampled by the on-line monitoring system in one day.

We also compared the thermocouples according to their locations above the core. Figure 11.9 shows a simplified map of the core with the locations of the thermocouples. The core is divided into four quadrants as shown in the figure. Assuming that the water temperature around all thermocouples in each quadrant is the same, we have obtained the results shown in Figure 11.10 for groups of thermocouples in each of the four quadrants. The results are shown for both operating and shutdown conditions. The outliers are not shown in the figure. Again, it is apparent that the thermocouples are more accurate at low temperatures ($\sim$110°F) than high temperatures ($\sim$620°F). The results shown in Figure 11.10 correspond to the deviation of each thermocouple in a quadrant from the average of all thermocouples in the quadrant. In obtaining the average temperature of each quadrant, the outliers in the quadrant were excluded for the average based on a rejection criteria of $\pm$15°F.

In addition to the cross calibration tests described above, like signal comparison analyses of the core exit thermocouples were performed for each quadrant using about nine months of on-line monitoring data. The results are shown in Figure 11.11. Note that except for a few outliers in each quadrant, the deviations of thermocouples are well within about $\pm$9°F or approximately 1.5% of the reading, and that no significant drift is apparent in the data. The results in Figure 11.11 include a period when the plant was shut down for a few weeks. During this shutdown period, the

deviations of the outliers in each quadrant decreased to around zero indicating that the thermocouple errors are temperature dependent. This conclusion is consistent with common knowledge that faulty temperature sensors often show relatively small errors at low temperatures and larger errors at high temperatures. A plot of the deviations of four thermocouples as a function of temperature is shown in Figure 11.12. The group includes one good thermocouple (D07) and three bad thermocouples (J10, G12, and J02). Note that the deviation of the good thermocouple is independent of temperature, while that of the three bad thermocouples increase significantly as the temperature is increased. An additional plot of thermocouple errors versus temperature is shown in Figure 11.13 for three CETs. It is apparent that at about seven weeks into the data when the plant tripped, the deviations of the thermocouples decreased to near zero, especially that of the most faulty thermocouple in the group (number A06).

The reader should be reminded that all data plotted in the figures shown in this chapter have been window averaged to reduce noise. Therefore, sudden changes in the data may appear as transient changes.

The deviations of the McGuire hot leg RTDs are shown in Figure 11.14. These results represent the deviations of the individual RTDs from the average of all twelve hot leg RTDs; three in each of the plant's four loops. The same type of data is shown in Figure 11.15 for the four cold leg RTDs along with a plot of the temperature indication of the RTDs. No significant drift is apparent in the hot leg or cold leg RTD data for the nine months of monitoring. The reason for the large spikes in the cold leg RTDs is not known. It appears that the spikes are not from the RTDs. They could be due to surveillance tests during which the RTDs were temporarily removed from service. Except for the spikes, the cold leg RTDs have normal behavior and agree well with each other to within 1°F. This compares with 4°F for the hot leg RTDs. The reason that the hot leg RTDs have larger deviations is probably due to the temperature stratification which is a more significant problem in the hot legs than the cold legs. Figure 11.16 shows the deviation of one of the hot leg RTDs versus the hot leg temperature and reactor power. This data set shows that the deviation of the RTD increases from around zero to near 2.5°F as the reactor power is increased from zero to about 100 percent. This confirms that the hot leg RTD deviations are mostly due to temperature stratifications. Two traces are shown in each of the two graphs in Figure 11.16. One trace corresponds to the RTD deviation as the power is

Table 11.2

Results of Like Signal Comparison for
Core Exit Thermocouples at McGuire Unit 2

| Item | Tag Number | Deviation (°F) | |
		Full Power	Shutdown
1	D03	0.479	-1.003
2	F05	1.675	-0.069
3	A06	-20.862	-2.049
4	C04	0.234	-0.568
5	D07	-1.367	0.640
6	E02	-0.258	-0.128
7	G04	7.596	2.154
8	C08	3.437	-0.757
9	D11	1.564	0.075
10	F09	3.726	-0.976
11	F15	8.475	1.685
12	G12	-61.356	-0.647
13	B11	-1.128	0.081
14	D13	0.817	-0.868
15	E10	-1.336	-1.168
16	G14	1.804	-0.736
17	J10	-64.004	-0.312
18	K13	4.522	0.555
19	L08	2.116	0.779
20	M11	4.970	0.955
21	M13	3.683	-0.768
22	P09	0.342	-0.786
23	H11	6.412	1.146
24	J08	5.477	1.169
25	K15	-21.315	-0.806
26	L12	5.597	2.389
27	N10	3.185	1.688
28	N14	-37.898	-0.985
29	R08	-18.651	-1.976
30	K01	-16.692	-2.226
31	K05	3.078	0.613
32	N02	-35.023	-2.197
33	P05	0.283	-0.004
34	J02	-62.557	0.561
35	L02	5.265	1.096
36	N04	2.585	2.457
37	N06	4.557	0.986
Average Temperature (°F)		623.5	112.0

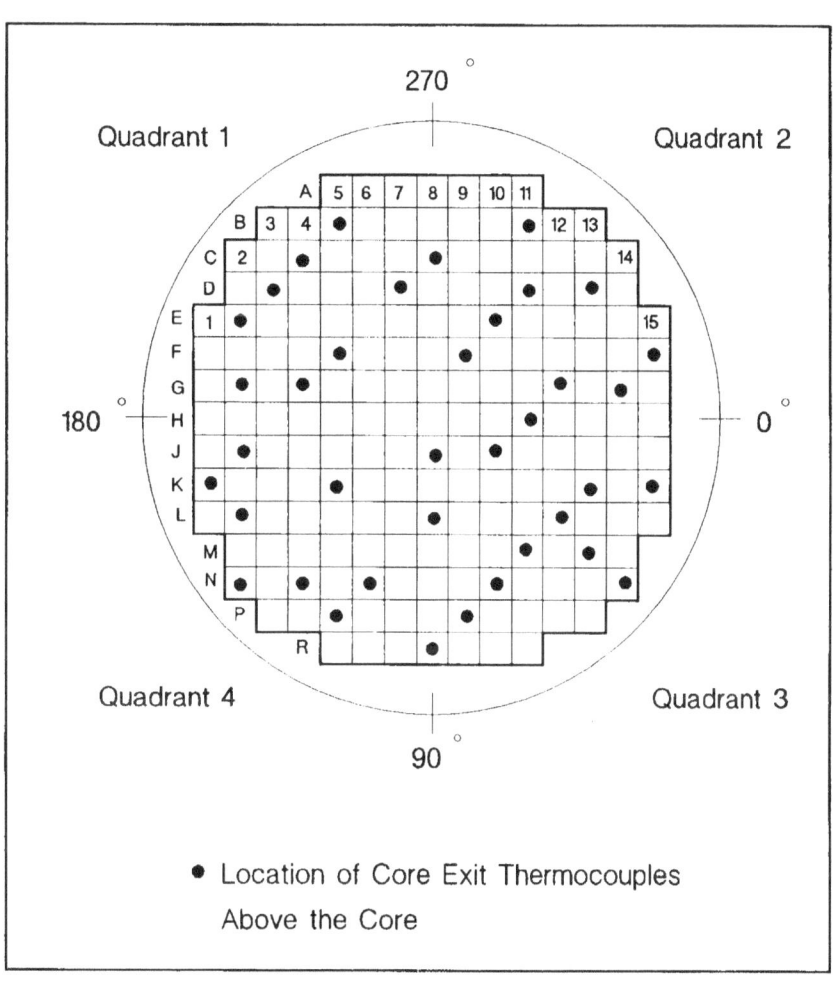

Figure 11.9   McGuire core exit thermocouple map

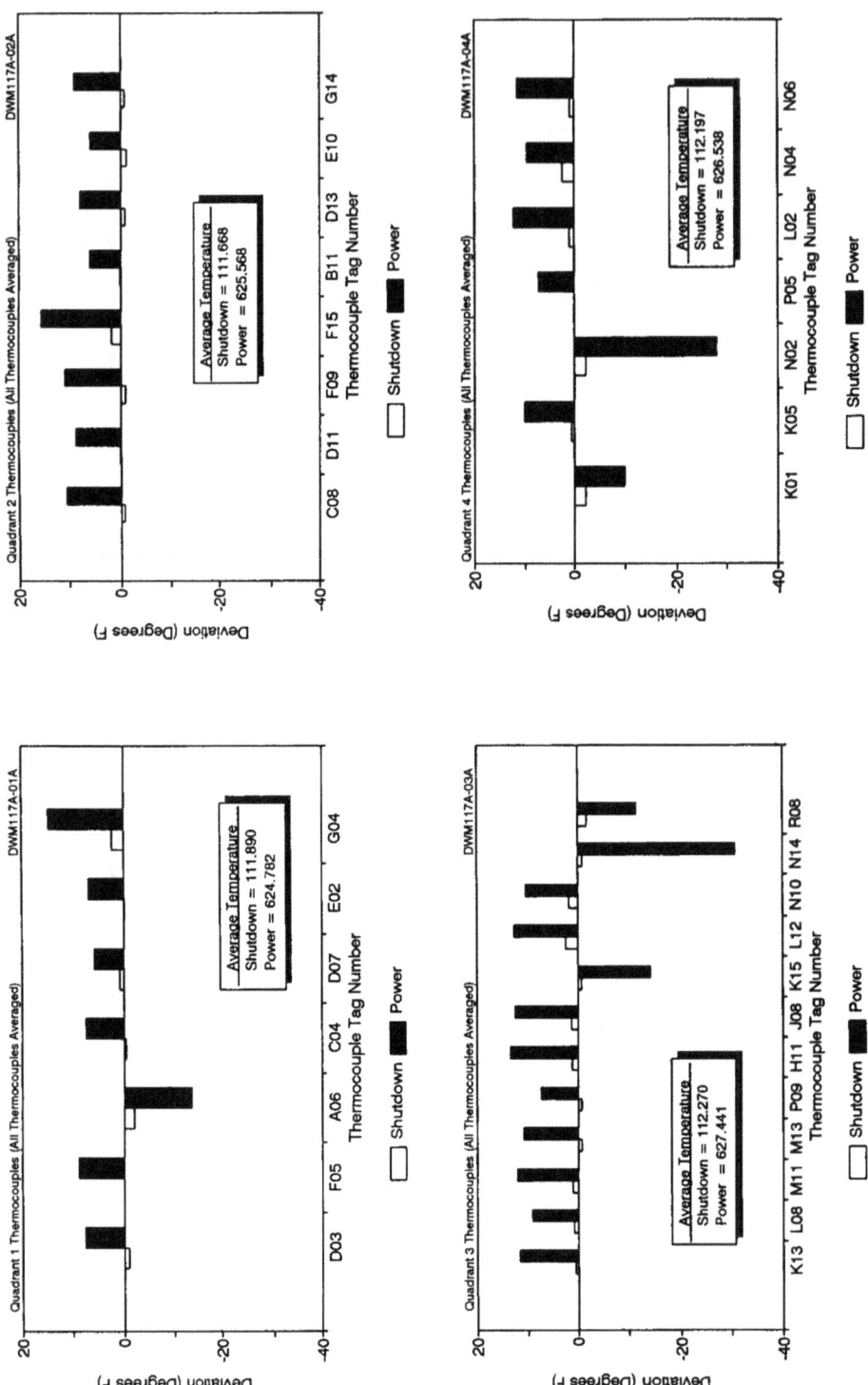

Figure 11.10 Deviation of CETs in each quadrant

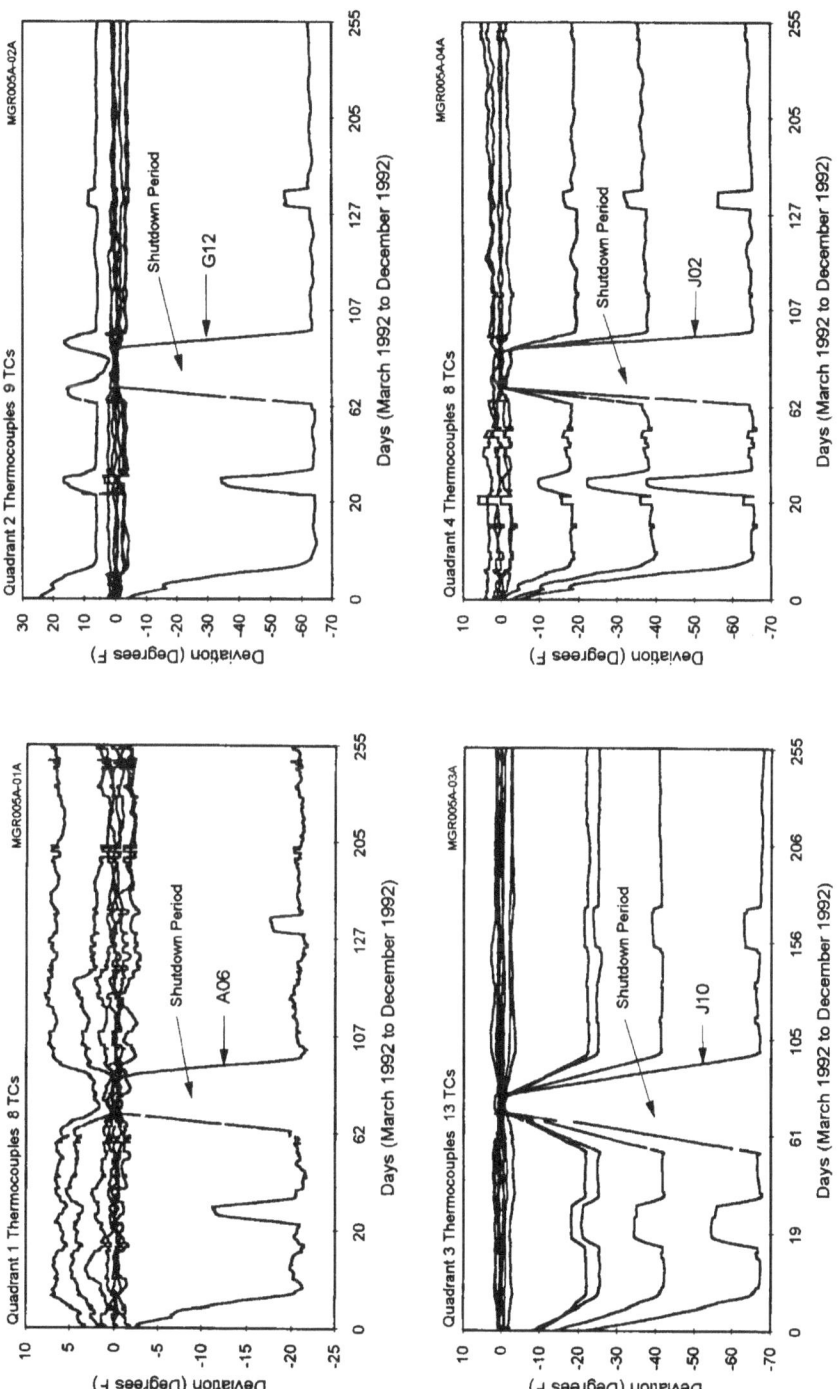

Figure 11.11 On-line monitoring results for core exit thermocouples

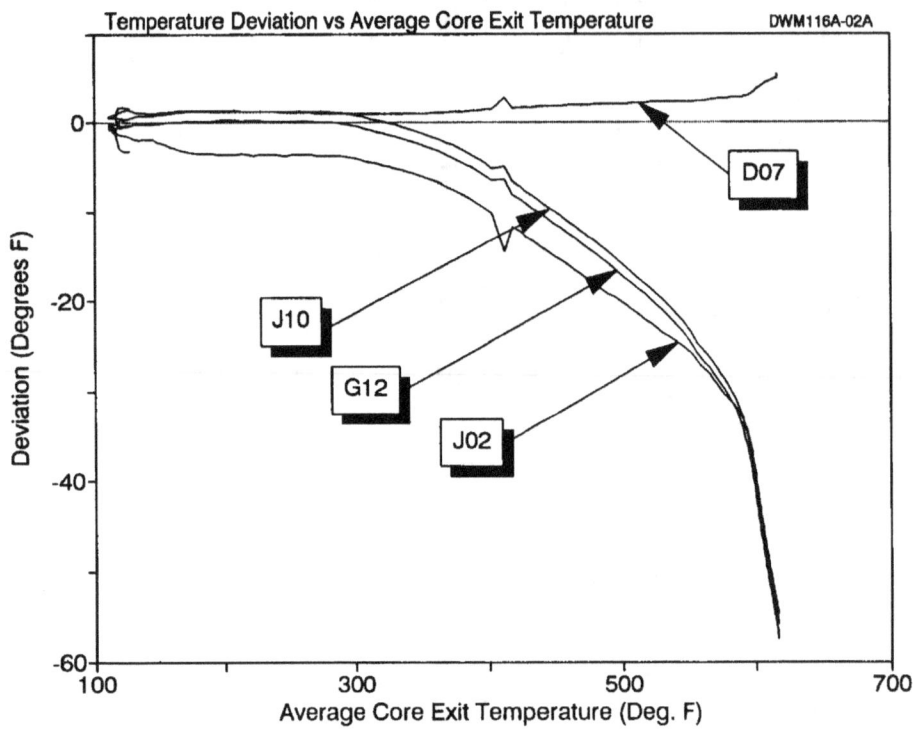

Figure 11.12   Deviation of selected thermocouples as a function of temperature

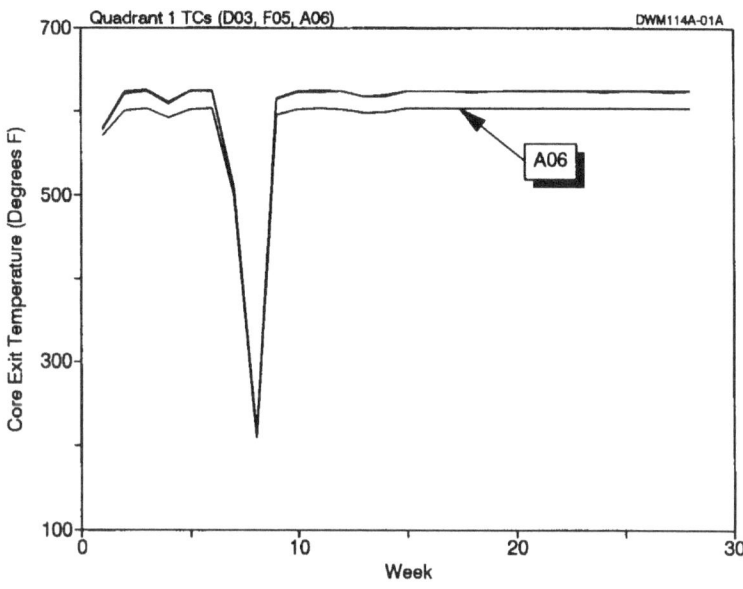

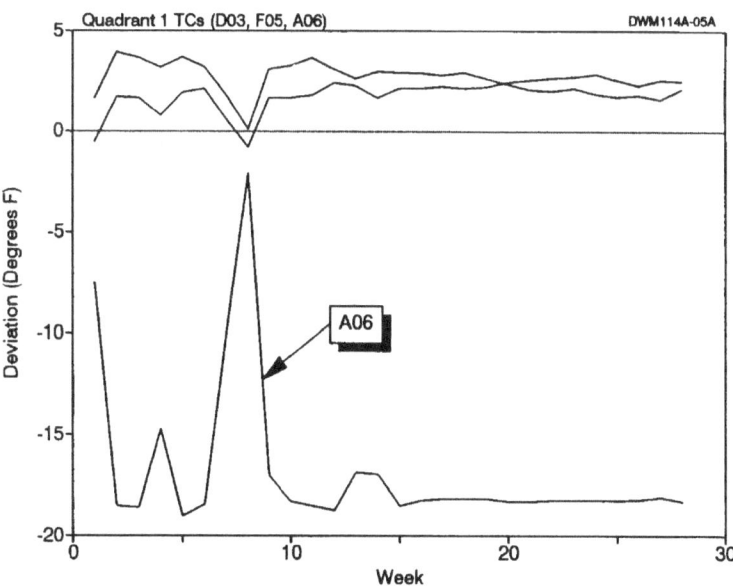

Figure 11.13 Normal reading and deviation of selected thermocouples before, during and after a plant trip

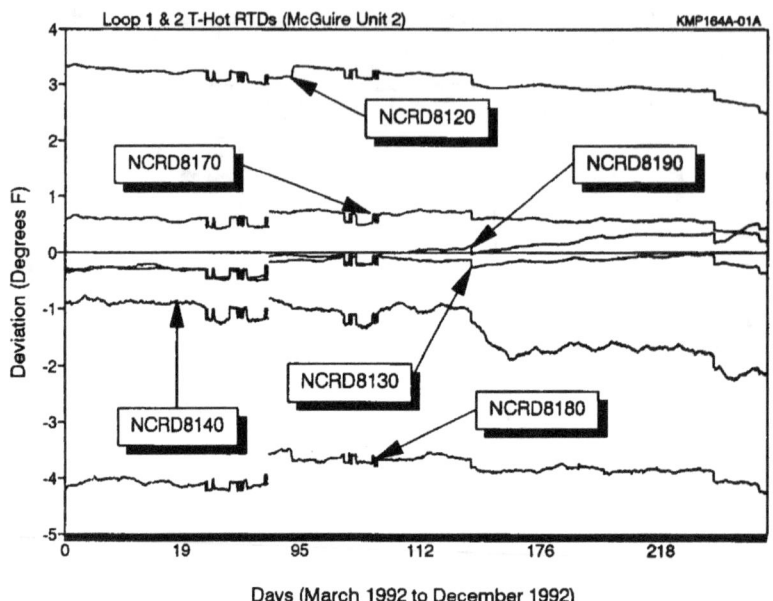

Days (March 1992 to December 1992)

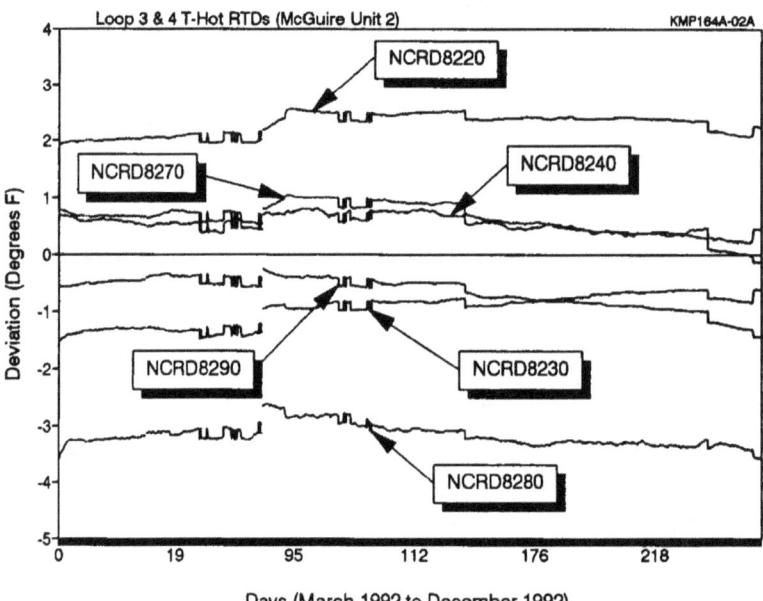

Days (March 1992 to December 1992)

Figure 11.14   On-line monitoring results for hot leg RTDs

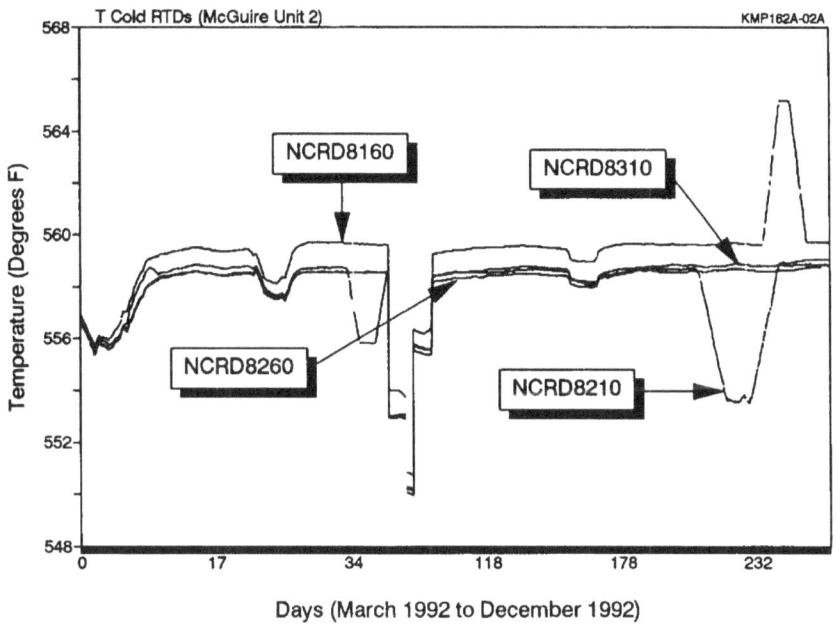

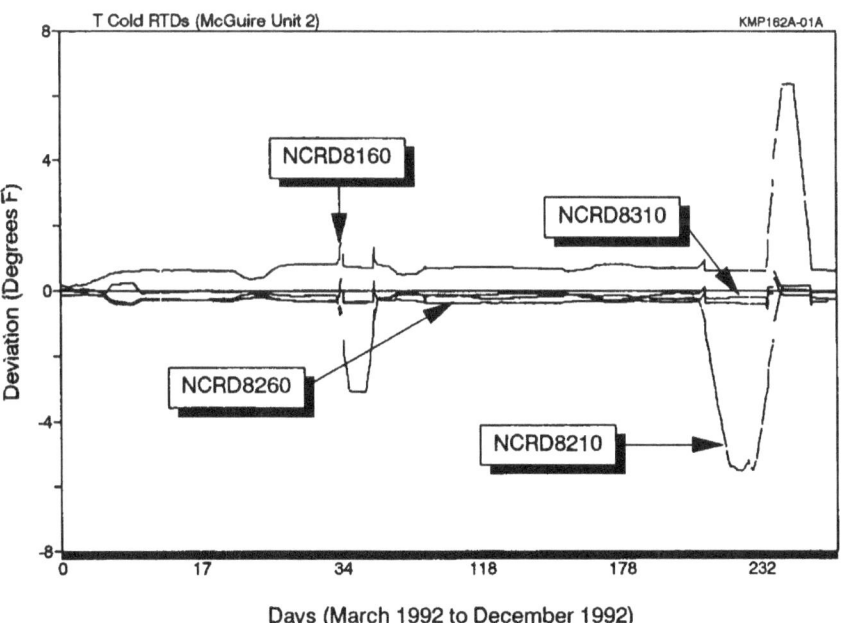

Figure 11.15   On-line monitoring results for cold leg RTDs

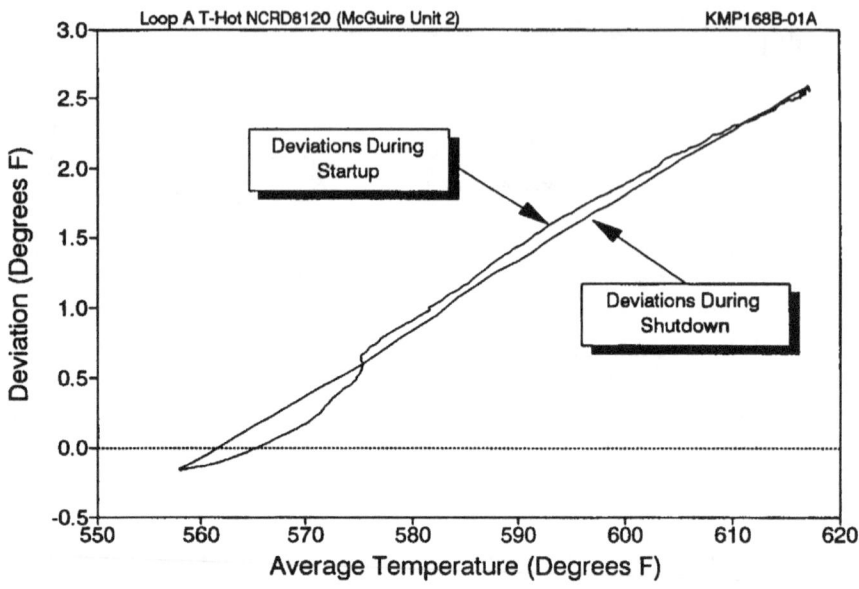

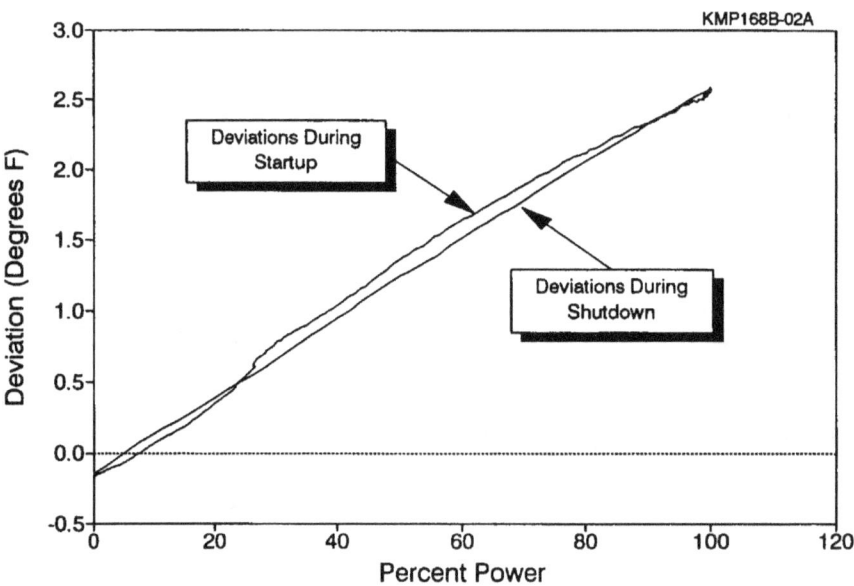

Figure 11.16   Temperature stratification errors as a function of hot leg
temperature (top) and reactor power (bottom)

increased during a startup, and another trace is for the RTD deviation during a shutdown. The two traces are within about 0.1°F of one another which is a very good agreement considering the uncertainties in this type of measurement.

Figure 11.17 shows the deviations of all hot leg RTDs in loop A as a function of reactor power. Note that only one of the three RTDs shows a significant deviation as the power is increased. This is the same RTD that was shown in Figure 11.16. In Figure 11.16 the deviation of the NCRD8120 is plotted versus the average of all three hot leg RTDs, and in Figure 11.17, the deviation of the NCRD8120 is plotted versus the average of all three RTDs until about 40 percent power when the deviation of NCRD8120 exceeds the consistency criteria of 1°F. At this point NCRD8120 is excluded from the average as evident by the sudden shifts in the levels of the three deviations. As a result, this RTD shows approximately 4°F of temperature stratification as opposed to about 2.5°F when all three signals were averaged together.

The same type of data shown above for thermocouples and RTDs are available for the neutron detectors (also referred to as Nuclear Instrumentation or NI channels). Figure 11.18 shows like signal comparison results for the NI signals. Two plots are shown in Figure 11.18, each with four NI signals. The deviations of each signal in the two plots is calculated from the average of the four signals in the group. These results show that the NI signals agree with each other within about ±0.5 percent throughout the monitoring period and no significant drift is apparent in the data.

Figures 11.19 through 11.22 show representative deviation plots for some of the services being monitored at McGuire that have not been discussed earlier in the report. The results for each group of signals represent the deviations of the individual signals from the average of the signals in the group. For those plots where there are only two signals in the group (Figures 11.21 and 11.22), the deviations appear as mirror images of one another as expected.

In addition to the DC data analyses described above, we performed noise analysis on a large group of McGuire signals. However, since the main focus of this project is on calibration drift monitoring, we will include only a few examples of AC data analysis. The AC results are included in terms of PSD plots from FFT analysis. Figure 11.23 shows six PSDs for a steam generator level signal at various plant conditions. It is obvious that there is no noise signal at cold shutdown (trace 6) and the zero power data (trace 5) does not have good resolution. However, the data at power is very good and repeatable as indicated by the four PSDs that are superimposed on each other. The same type of results are shown in Figure 11.24 for a hot leg RTD. This is followed by Figure 11.25 comparing the PSDs of a steam generator level signal for 100 percent and 18 percent power levels. Except for a resonance in the data for the 18 percent power, the PSD results are identical in terms of the information they contain about the dynamics of the transmitter.

In Figure 11.26 two PSD plots for a steam generator level transmitter are shown. The first PSD is from noise data obtained in December 1991 for response time testing of this sensor, and the second PSD is from noise data acquired by the on-line monitoring system in September 1992. Two points are noteworthy: 1) the two PSDs are very similar in shape and have the same roll-off characteristics meaning that the response time of this transmitter did not change between 1991 and 1992, and 2) the quality of PSDs is independent of the data acquisition method, i.e., the December 1991 data which used optimum data collection equipment and the September 1992 data which used a simple noise monitoring technique have produced the same results. That is, when the amplitude of noise in the process signal is adequate, a simple data acquisition procedure such as the one used in the on-line monitoring system should be adequate for monitoring for response time changes.

Figure 11.27 shows raw DC data and corresponding PSDs for three hot leg RTDs at McGuire. As expected, the low frequency amplitude of the PSDs correspond to the amplitude of the noise in the DC data. This low frequency amplitude provides diagnostic information about the sensor and the process and can be tracked to determine changes in the process or the sensor. Another useful characteristic of these PSDs is their roll-off rate and roll-off frequency that can be monitored to detect response time changes.

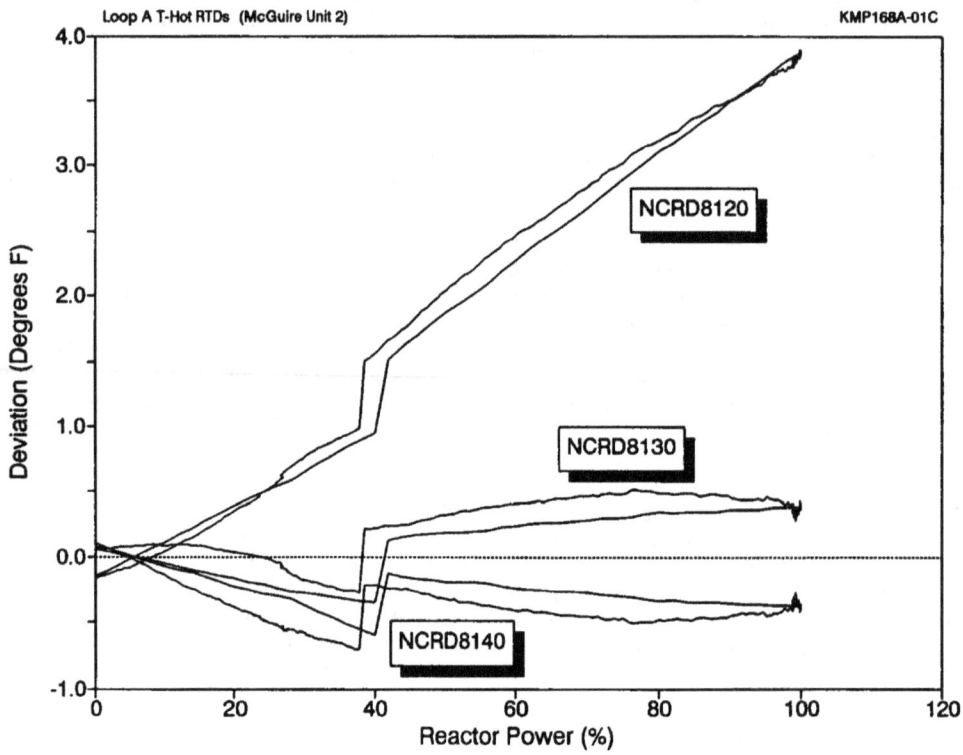

Figure 11.17   Deviation of hot leg RTDs in loop A

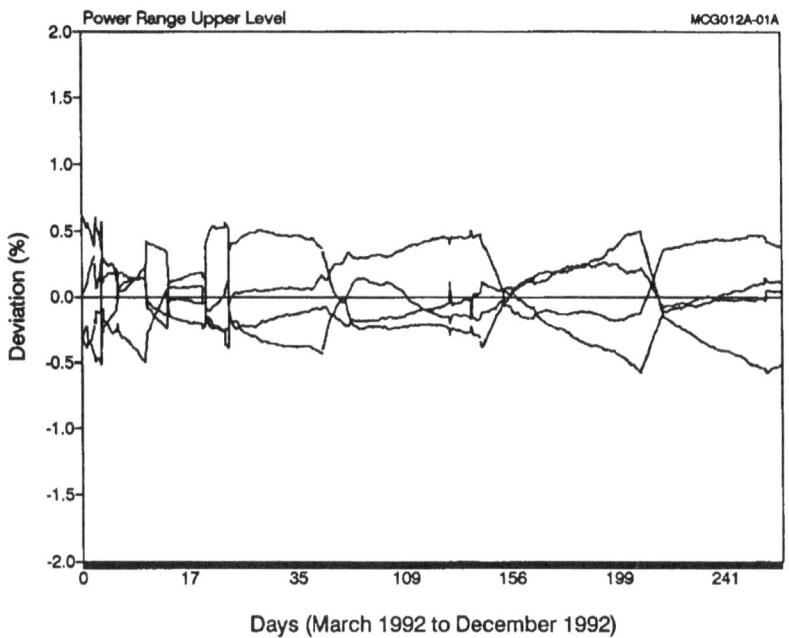

Days (March 1992 to December 1992)

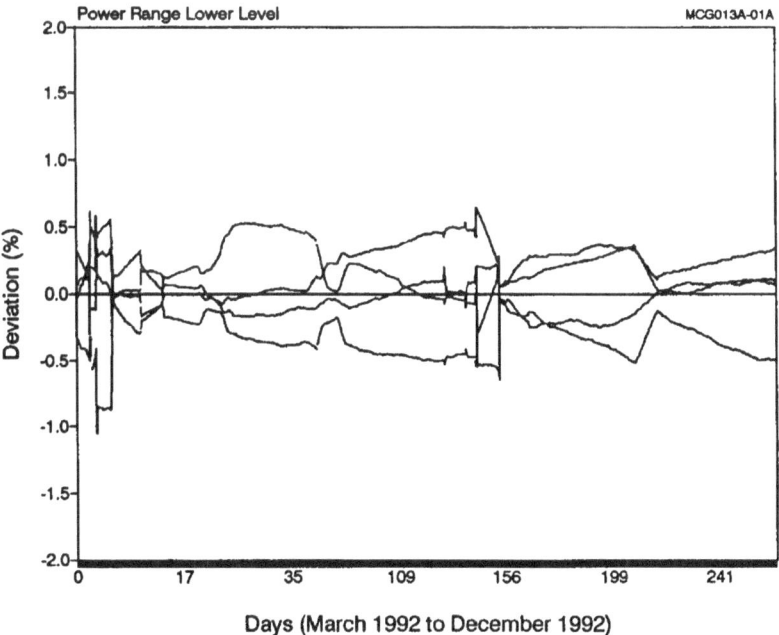

Days (March 1992 to December 1992)

Figure 11.18   On-line monitoring results for NI channels

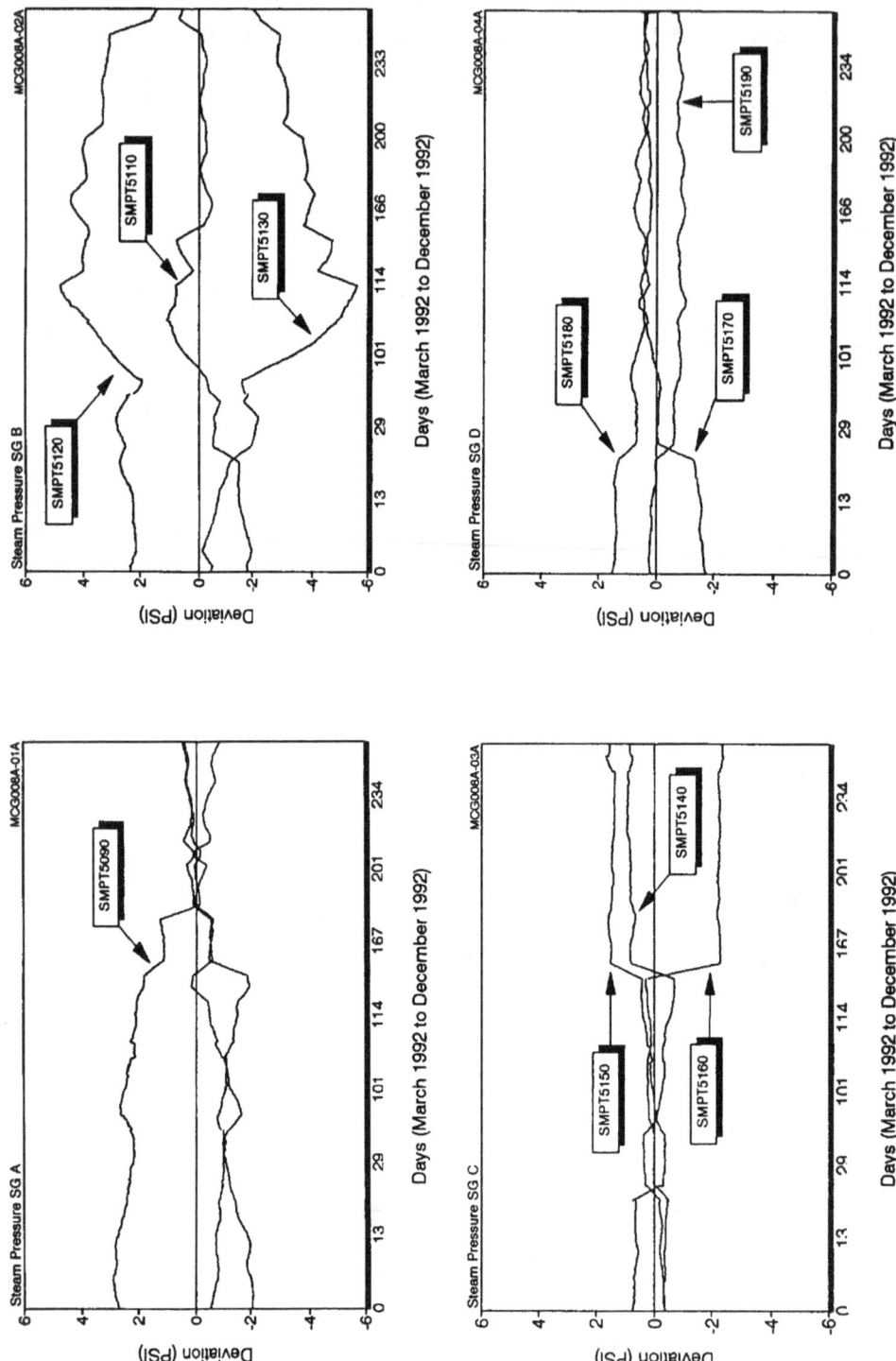

Figure 11.19 On-line monitoring results for steam pressure for four steam generators

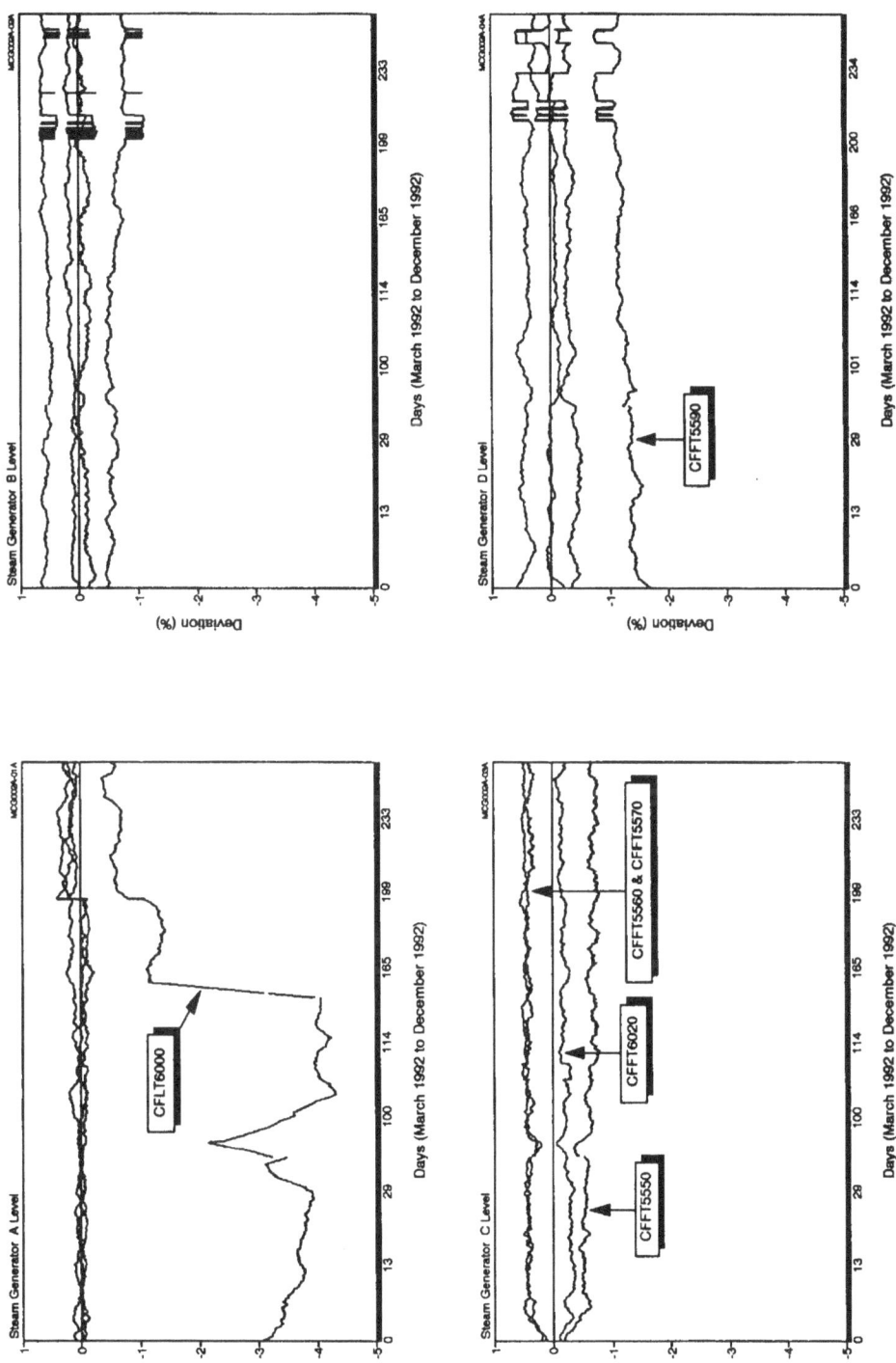

Figure 11.20   On-line monitoring results for level signals for four steam generators

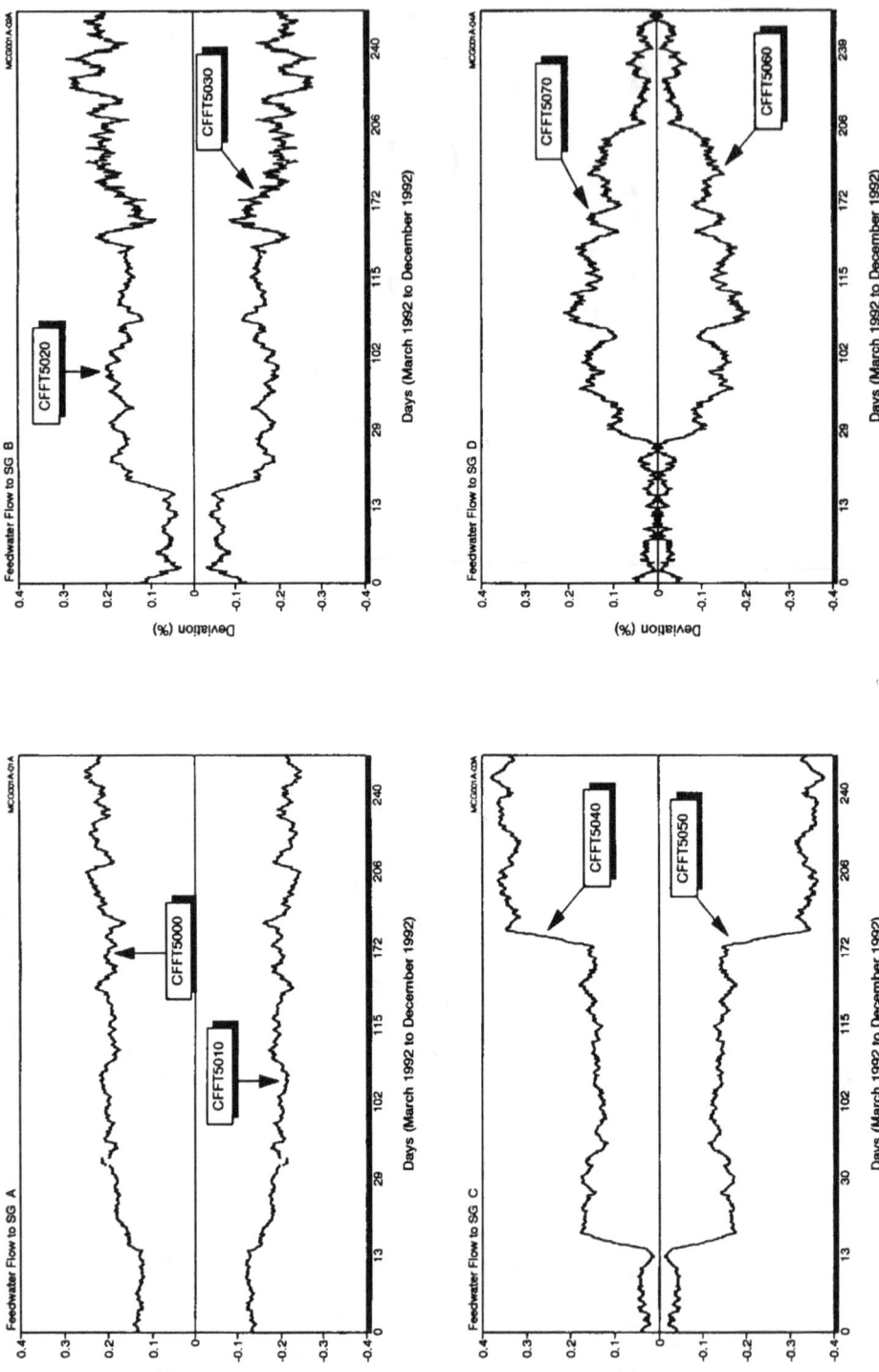

Figure 11.21 On-line monitoring results for feedwater flow signals for four steam generators

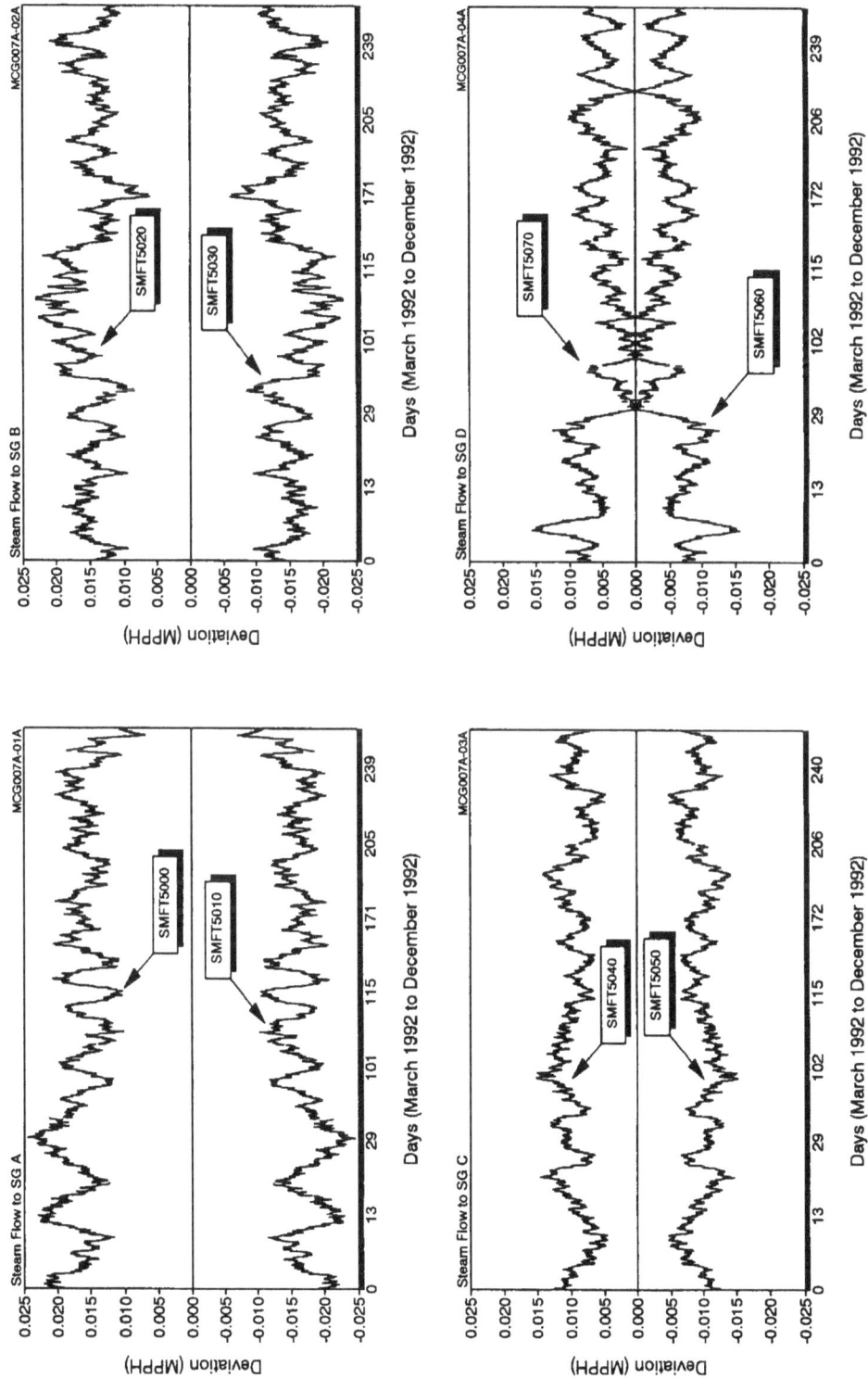

Figure 11.22 On-line monitoring results for steam flow signals for four steam generators

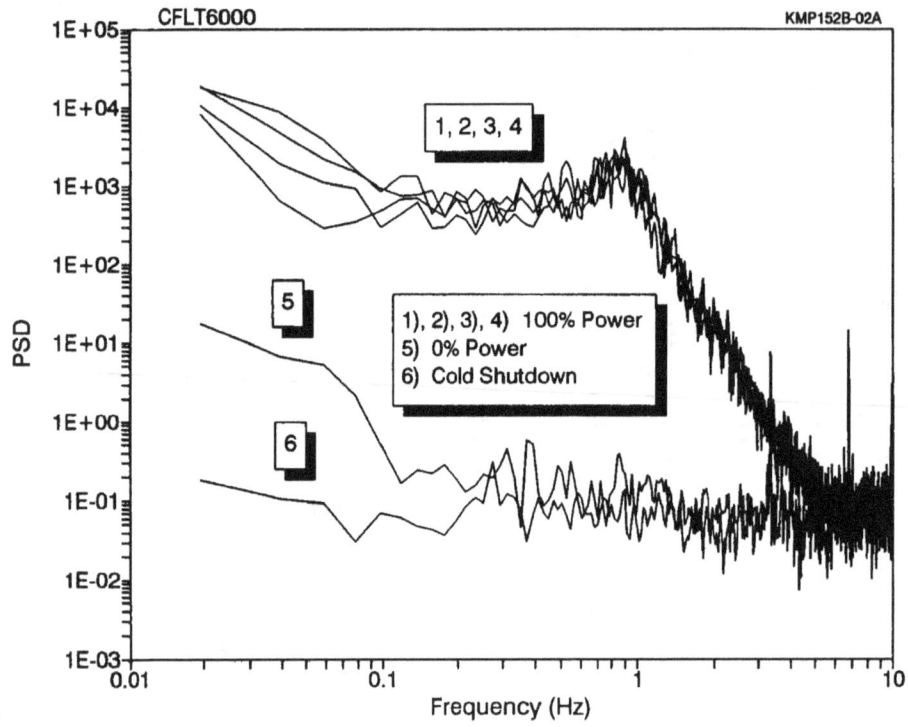

Figure 11.23   PSD results as a function of operating conditions for a level transmitter

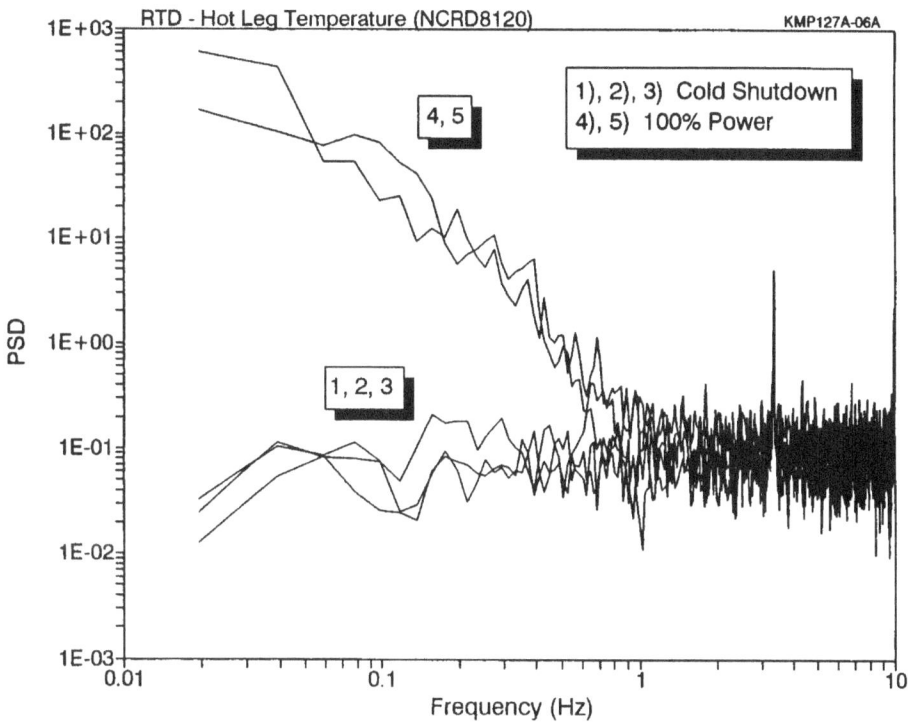

Figure 11.24   PSD results as a function of operating conditions for a hot leg RTD

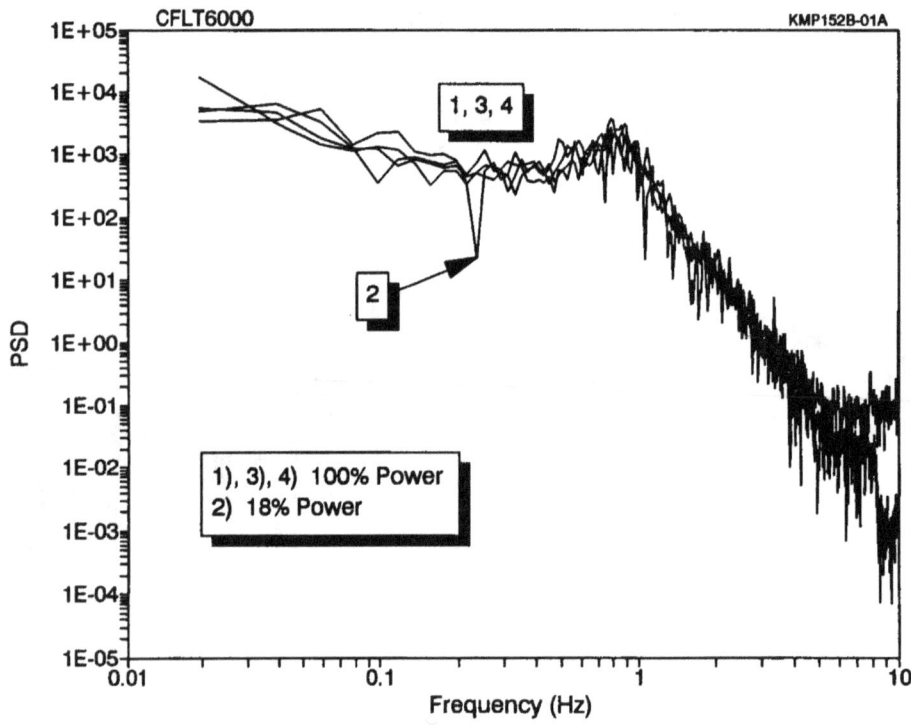

Figure 11.25   PSD results for a level transmitter at 18 and
100 percent power

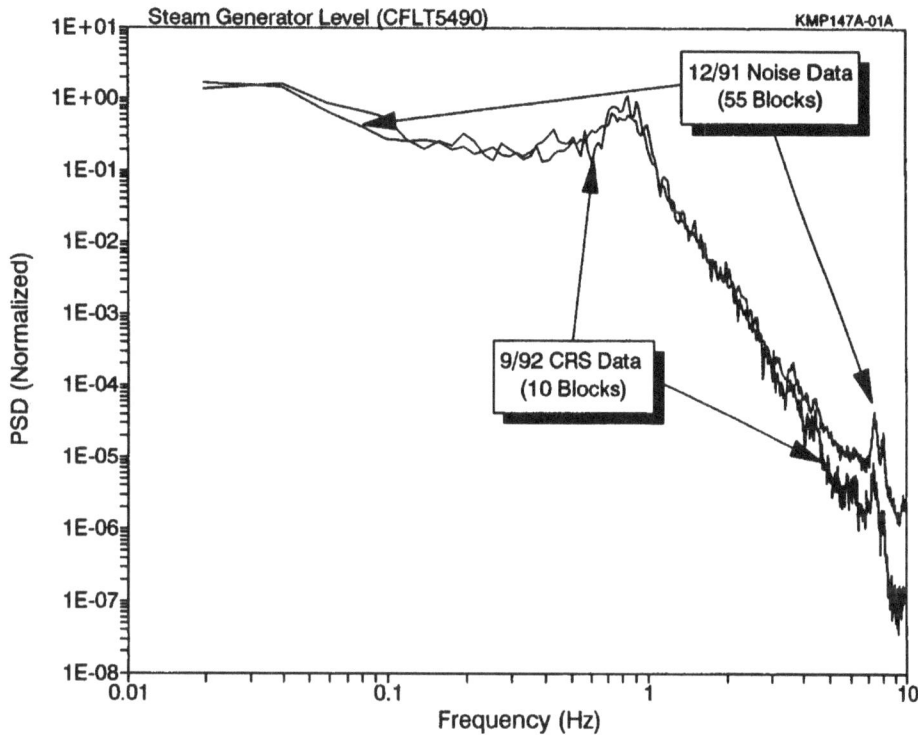

Figure 11.26   Comparison of PSD results for optimum data acquisition and simple data acquisition

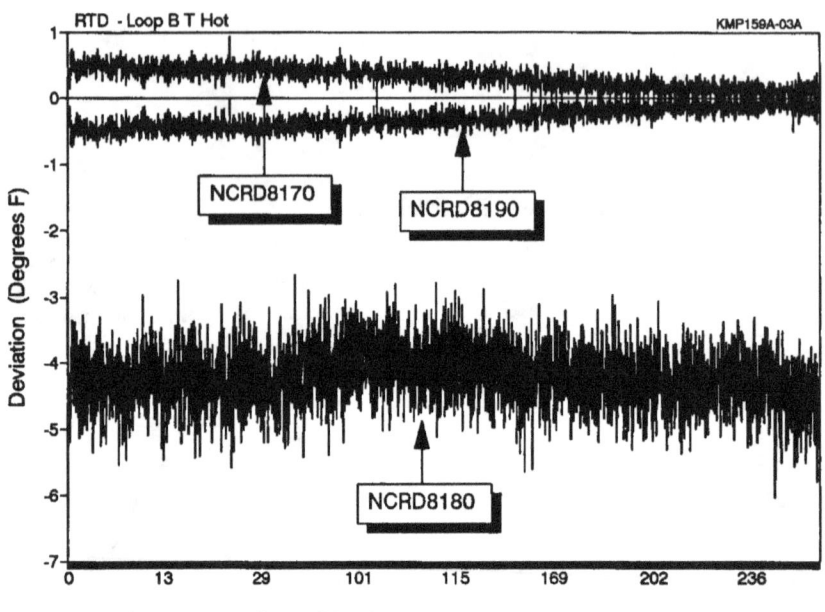

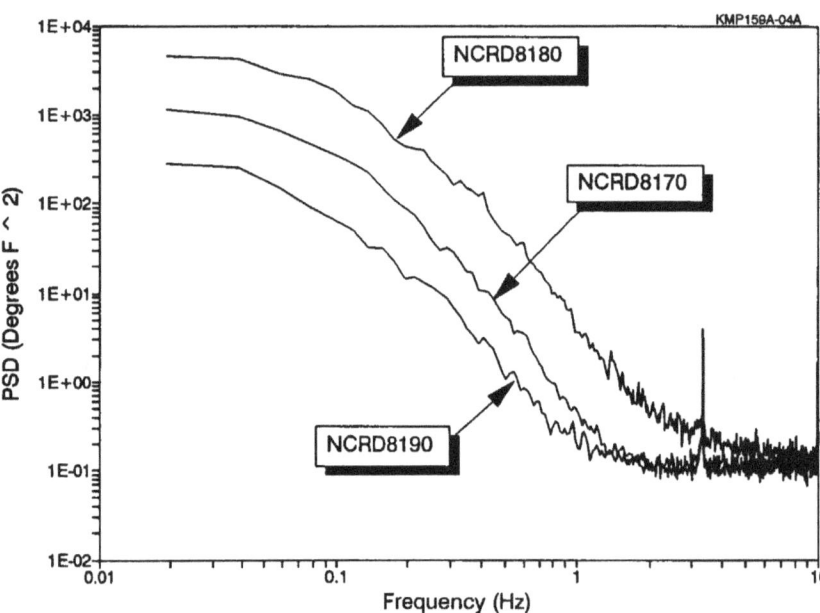

Figure 11.27   Raw data and corresponding PSDs for three hot leg RTDs

12. Conclusion

A feasibility study was successfully completed on the validity of on-line monitoring techniques for remote testing of calibration and response time of process instrumentation channels in nuclear power plants. This work involved research with nuclear grade temperature and pressure instrumentation in simulated reactor conditions in a laboratory, and in-plant validation work at the McGuire Nuclear Power Station, a four-loop PWR.

The project focused on DC data analysis for drift monitoring. In addition, limited work was performed on AC data analysis for response time degradation testing.

For the DC data analysis, simple methods such as straight and weighted averaging of redundant signals and more complicated methods such as empirical and physical modeling were researched. For AC signal analysis, Fast Fourier Transform, autoregressive modeling, and zero crossing techniques were examined using both laboratory and in-plant data. These efforts have successfully laid the foundation for an in-depth study to quantify the accuracy and reliability of the on-line monitoring techniques for instrument calibration reduction and response time degradation monitoring in nuclear power plants. The in-depth study is currently underway in a Phase II project to be completed in 1994.

References

1. Kujawski, E., Jacobs, I. M., Smith, A. M., EPRI NP-5081, "An Evaluation of the Use of Signal Validation Techniques as a Defense Against Common-Cause Failures," Electric Power Research Institute, Palo Alto, California, February 1987.

2. Swisher, V. I., EPRI NP-5389, "User's Guide for Signal Validation Software," Electric Power Research Institute, Palo Alto, California, September 1987.

3. NRC Meeting on Instrument Calibration Reduction, Washington, D.C., December 1992.

4. ISA Standard S67.06, "Response Time Testing of Nuclear Safety-Related Instrument Channels in Nuclear Power Plants," Instrument Society of America, Research Triangle Park, North Carolina, 1984.

5. Meijer, C. H., Pasquenza, J. P., EPRI NP-2110, "On-Line Power Plant Signal Validation Technique Utilizing Parity-Space Representation and Analytical Redundancy," Electric Power Research Institute, Palo Alto, California, November 1981.

6. ECAD Trends, The Newsletter for Protective Maintenance of Electrical Systems and Components, Vol. 3, No. 1, ECAD Division of CM Technologies Corporation, Coraopolis, Pennsylvania, July 1992.

7. The Journal of the National Academy for Nuclear Training, "In the Loop," The Nuclear Professional Magazine, Summer 1992.

8. Hashemian, H. M., et al., NUREG/CR-5560, "Aging of Nuclear Plant Resistance Temperature Detectors," U.S. Nuclear Regulatory Commission, Washington, D.C., June 1990.

9. Shepard, R. L., EPRI NP-5537, "Remote Calibration of Resistance Temperature Devices," Electric Power Research Institute, Palo Alto, California, February 1988.

10. Hashemian, H. M., et.al., NUREG/CR-5383, "Effect of Aging on Response Time of Nuclear Plant Pressure Sensors," U.S. Nuclear Regulatory Commission, Washington, D.C., June 1989.

11. Upadhyaya, B. R., Holbert, K. E., Eryurek, E., Proceedings of the 35th International Instrumentation Symposium, ISA 89-0026, "Automated Generation of Non-Linear System Characterization for Sensor Failure Detection," Instrument Society of America, Orlando, Florida, May 1989.

12. Holbert, K. E., Ph.D. Dissertation, "Comprehensive Signal Validation for Nuclear Power Plants," University of Tennessee, Knoxville, Tennessee, August 1989.

13. Hashemian, H. M., et. al., "Apparatus for Measuring the Degradation of a Sensor Time Constant," United States Patent Number 4295128, October 1981.

14. Nuclear Management and Resources Council, Report #91-02, "Summary Report of NUMARC Activities to Address Oil Loss in Rosemount Transmitters," April 1991.

NRC FORM 335
(2-89)
NRCM 1102,
3201, 3202

U.S. NUCLEAR REGULATORY COMMISSION

BIBLIOGRAPHIC DATA SHEET

(See instructions on the reverse)

1. REPORT NUMBER
(Assigned by NRC. Add Vol., Supp., Rev., and Addendum Numbers, if any.)

NUREG/CR-5903

2. TITLE AND SUBTITLE

Validation of Smart Sensor Technologies for Instrument Calibration Reduction in Nuclear Power Plants

3. DATE REPORT PUBLISHED

MONTH	YEAR
January	1993

4. FIN OR GRANT NUMBER

L2010

5. AUTHOR(S)

H. M. Hashemian, D. W. Mitchell, K. M. Petersen, C. S. Shell

6. TYPE OF REPORT

Technical

7. PERIOD COVERED *(Inclusive Dates)*

9/30/91 - 12/31/92

8. PERFORMING ORGANIZATION – NAME AND ADDRESS *(If NRC, provide Division, Office or Region, U.S. Nuclear Regulatory Commission, and mailing address; if contractor, provide name and mailing address.)*

Analysis and Measurement Services Corporation
AMS 9111 Cross Park Drive NW
Knoxville, TN 37923-4599

9. SPONSORING ORGANIZATION – NAME AND ADDRESS *(If NRC, type "Same as above"; if contractor, provide NRC Division, Office or Region, U.S. Nuclear Regulatory Commission, and mailing address.)*

Division of Engineering
Office of Nuclear Regulatory Research
U. S. Nuclear Regulatory Commission
Washington, D. C. 20555

10. SUPPLEMENTARY NOTES

Copyrighted by Analysis and Measurement Services Corporation, 1993

11. ABSTRACT *(200 words or less)*

This report presents the preliminary results of a research and development project on the validation of new techniques for on-line testing of calibration drift of process instrumentation channels in nuclear power plants. These techniques generally involve a computer-based data acquisition and data analysis system to trend the output of a large number of instrument channels and identify the channels that have drifted out of tolerance. This helps limit the calibration effort to those channels which need the calibration, as opposed to the current nuclear industry practice of calibrating essentially all the safety-related instrument channels at every refueling outage.

12. KEY WORDS/DESCRIPTORS *(List words or phrases that will assist researchers in locating the report.)*

Calibration
Calibration Reduction
Cross Calibration
Drift
In-Situ Testing
Maintenance
Nuclear Power Plants
On-Line Monitoring
On-Line Testing

Process Instrumentation Channels
Response Time Testing
Signal Validation
Smart Sensors
Smart Sensor Technologies

13. AVAILABILITY STATEMENT

Unlimited

14. SECURITY CLASSIFICATION

(This Page)

Unclassified

(This Report)

Unclassified

15. NUMBER OF PAGES

16. PRICE

NRC FORM 335 (2-89)

Printed on recycled paper

Federal Recycling Program

www.ingramcontent.com/pod-product-compliance
Lightning Source LLC
Chambersburg PA
CBHW080248180526
45167CB00006B/2455